Ergodic Theory
and Fractal Geometry

Conference Board of the Mathematical Sciences

CBMS

Regional Conference Series in Mathematics

Number 120

Ergodic Theory and Fractal Geometry

Hillel Furstenberg

Published for the
Conference Board of the Mathematical Sciences
by the
American Mathematical Society
Providence, Rhode Island
with support from the
National Science Foundation

NSF/CBMS Regional Conference in the Mathematical Sciences on Ergodic
Methods in the Theory of Fractals, held at Kent State University,
June 18–23, 2011

2010 *Mathematics Subject Classification*. Primary 28A80, 37A30; Secondary 30D05,
37F45, 47A35.

For additional information and updates on this book, visit
www.ams.org/bookpages/cbms-120

Library of Congress Cataloging-in-Publication Data
Furstenberg, Harry.
    Ergodic theory and fractal geometry / Hillel Furstenberg.
        pages cm. — (Conference Board of the Mathematical Sciences Regional Conference series
in mathematics ; number 120)
    "Support from the National Science Foundation."
    "NSF-CBMS Regional Conference in the Mathematical Sciences on Ergodic Methods in the
Theory of Fractals, held at Kent State University, June 18–23, 2011."
    Includes bibliographical references and index.
    ISBN 978-1-4704-1034-6 (alk. paper)
    1. Ergodic theory—Congresses. 2. Fractals—Congresses. I. Conference Board of the Math-
ematical Sciences. II. National Science Foundation (U.S.) III. Title.

QA313.F87   2014
515'.48—dc23
                                                                                2014010556

Dedicated to the memory of Benoit Mandelbrot

Contents

Preface

Dynamics in all its variations is the study of change. In the usual physical context, change takes place within time. The objects of geometry are static and if there is any change, it is "in the eye of the beholder". In fractal geometry this point takes on meaning, particularly in the form of changing degree of magnification and "zooming in" on an object. This suggests developing dynamical concepts appropriate to this framework.

In these notes, based on a series of lectures delivered at Kent State University in 2011, we show that ergodic theoretic concepts can be applied to the process of changing magnification to give insight to phenomena peculiar to fractals. An important step is showing how fractal dimension can be interpreted in terms of ergodic averages in an appropriate measure preserving system. The familiar phenomenon of self similarity appears as the analogue of periodicity in classical dynamics. We don't pursue the full implications of recurrence in the geometric context, but some examples of the related Ramsey type questions are considered.

The theory developed here and the major ideas originated in the papers [**F**] and [**F′**]. It will develop that there is a close connection between dimension theory and rates of growth of trees. This is exploited in [**FW**] where analogs of Szemerédi's theorem are demonstrated in the context of trees.

I am indebted to Dmitry Ryabogin and Fedor Nazarov for transcribing the lectures as well as for working out many details that were not provided in the lectures as I presented them.

Hillel Furstenberg, January, 2014
Jerusalem, Israel

CHAPTER 1

Introduction to Fractals

Unlike many other mathematical notions, the notion of a fractal is not really well-defined by some axiomatic list of properties. The best way to understand what we mean by a fractal is to contrast fractals with classical geometry objects like lines and surfaces. The latter are usually studied using tools from Differential Geometry and, ultimately, Linear Algebra. They are *linear* objects, which are locally well approximated by their tangent planes. This linearity is absent when we deal with fractals.

To formalize this concept of non-linearity , we will introduce the notions of *mini-* and *micro-sets*. We will start with the definition of the Hausdorff distance between subsets of a metric space.

DEFINITION. *Let X be a compact metric space and let $\sigma(X)$ be the set of all its closed subsets. If $x \in X$ and $A \in \sigma(X)$, we define the distance from x to A as the minimum of all distances from x to points of A, i.e.,*

$$d(x, A) = \min_{y \in A} d(x, y).$$

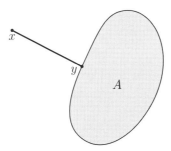

FIGURE 1. The distance from a point x to a set A is the distance from x to a nearest point $y \in A$.

If A, $B \in \sigma(X)$ are two closed subsets of X, we define the distance between them as

$$\mathcal{D}(A, B) = \max\{\max_{x \in A} d(x, B), \max_{y \in B} d(y, A)\}.$$

Thus, the inequality $\mathcal{D}(A, B) < \varepsilon$ means that A is contained in the ε-neighborhood of B and B is contained in the ε-neighborhood of A, i.e., for every $x \in A$, there exists $y \in B$ such that $d(x, y) < \varepsilon$ and vice versa.

Note that the sets close in this metric may be of very different nature from the topological perspective. For instance, the interval $[0, 1]$ is ε-close to the finite set of points $\{0, \frac{1}{n}, \frac{2}{n}, \ldots, 1\}$ if $n > \frac{1}{\varepsilon}$ despite the fact that $[0, 1]$ is a continuum and $\{0, \frac{1}{n}, \frac{2}{n}, \ldots, 1\}$ is a discrete set.

Now we are ready to define mini- and micro-sets. Note that this terminology is not universally accepted, so some other books may use different names for the same objects. We will always assume that our starting set A is a compact subset of $\mathbb{R}^n$ contained in the unit cube $Q = [0, 1]^n$.

DEFINITION. *A mini-set of A is any set of the kind $(\lambda A + u) \cap Q$ where $\lambda \geq 1$ and $u \in \mathbb{R}^n$ are such that $Q \subset \lambda Q + u$.*

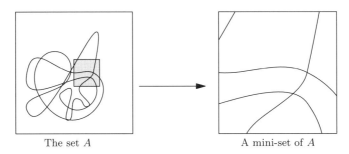

The set A A mini-set of A

FIGURE 2. A mini-set of A is the scaled intersection of A with a small square window (enlarged on the right).

A micro-set of A is the limit of any sequence A_n of mini-sets of A in the Hausdorff metric.

Informally, a mini-set is what you see if you look at a small portion of a set through a magnifying glass. The condition $Q \subset \lambda Q + u$ is needed to ensure that a mini-set of a mini-set of A is again a mini-set of A. Note also that the scales λ_n and the shifts u_n corresponding to the mini-sets A_n in the definition of a micro-set do not need to be related in any way.[1]

A lot of interesting micro-sets can be obtained by zooming at a single point like at the picture below, but we also have an option of moving our window around

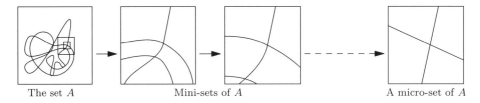

The set A Mini-sets of A A micro-set of A

FIGURE 3. A micro-set of a classical set is a union of several points.

simultaneously with the scale reduction.

The most interesting micro-sets are those that correspond to the sequences of mini-sets $A_n = (\lambda_n A + u_n) \cap Q$ with $\lambda_n \to +\infty$. For all objects studied in classical geometry, they are essentially flat, i.e., either planes or unions of a few planes as

---

[1]Note also that the set of closed subsets of the closed unit cube Q endowed with the Hausdorff metric is compact (Blaschke theorem), so any sequence of mini-sets contains a convergent subsequence.

in Figure 3. What distinguishes fractals from the classical geometry objects is that their micro-sets may be as complex as the original sets.

Consider, for example, the so-called *Sierpinski gasket S*.

FIGURE 4. Sierpinski gasket.

No matter how much you zoom in on it, it doesn't become flat or gets simpler in any other respect. More precisely, if you start zooming in at some fixed point of S, the corresponding family of mini-sets has periodic behavior: whatever you see at the scale λ, you see again at the scales 2λ, 4λ, 8λ, and so on (provided that the magnifying factor λ is large enough so that you can see only a small piece of S at once). For more complicated fractals the behavior can be more subtle, so that when zooming in one observes a chaotic behavior of shapes. The proper tools to handle this chaos are no longer those of Linear Algebra but those of Ergodic Theory.

Another difference between fractals and classical geometry objects is that fractals usually have non-integer (fractional) *dimension*. The classical notion of dimension is, essentially, a Linear Algebra notion. The dimension of a classical object like a line or a surface in $\mathbb{R}^n$ is nothing but the linear dimension of its tangent plane, so a point has dimension 0, a line has dimension 1, a surface has dimension 2, and so on. However, as a rule, fractals do not have tangent planes and this linear algebra approach becomes meaningless for them. The dimension of fractals has to be defined in a different way and can be any non-negative real number up to the dimension of the ambient Euclidean space.

We will return to this discussion in Chapter 2 and now we will look at how fractals arise in mathematics. The primary sources of fractal objects are infinite iterations of simple classical processes. Consider, for instance, the standard *middle-third Cantor set* on the line. Its construction starts with a closed interval, say $[0, 1]$:

C_0

FIGURE 5. The construction of the Cantor set starts with a single interval.

At the first step, the middle third $(\frac{1}{3}, \frac{2}{3})$ is removed and we get the set $C_1 = [0, \frac{1}{3}] \cup [\frac{2}{3}, 1]$:

C_1

FIGURE 6. At the first step the middle third is removed.

Next, the same procedure of removing the middle third is applied to each of the intervals of C_1 resulting in 4 intervals of length $\frac{1}{9}$:

C_2

FIGURE 7. This procedure is repeated for each remaining interval.

This process repeats again and again, so C_n consists of 2^n intervals of length 3^{-n}. The Cantor set C is defined as the limit of C_n in the Hausdorff metric, which in this case is the same as $\bigcap_{n=1}^{\infty} C_n$.

Note that the Lebesgue measure of C_n equals $(\frac{2}{3})^n$, which tends to 0 as $n \to \infty$, so we remove almost all points from the initial interval $[0, 1]$ in the sense of the Lebesgue measure. However, there are still many points left. For instance, all endpoints of the intervals appearing at all stages stay, so C is infinite. It is possible to show that C is of the cardinality of the continuum, so from this point of view, it is as large as the original interval $[0, 1]$.

The Sierpinski gasket S is just a 2-dimensional version of the middle third Cantor set. We start with with a closed equilateral triangle, split it into 4 equal triangles and remove the central one:

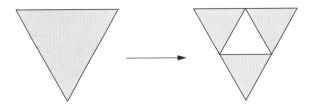

FIGURE 8. At the first step, the central triangle is removed.

Then we repeat this procedure for each of the three remaining triangles: and so on. After n steps, we get 3^n triangles with side-length 2^{-n}. Again, the area of S_n decays as a geometric progression. However, the total length $(\frac{3}{2})^n$ of the sides

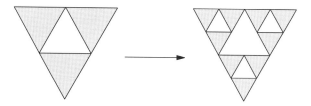

FIGURE 9. The procedure is repeated for each of the remaining triangles.

of the triangles constituting S_n grows exponentially. Thus, S is far too small for a 2-dimensional object and far too large for a 1-dimensional one, so its dimension has to be strictly between 1 and 2.

Another way to construct a fractal set is to start with some simple continuous map $f : \mathbb{R}^n \to \mathbb{R}^n$ and look at the set of points that do not escape to infinity under any number of iterations, i.e., the set of points for which the sequence x, $f(x)$, $f^2(x)$, ... remains bounded (here and below $f^n(x) = f(f(\ldots(f(x))\ldots))$ is the n'th iterate of f).

Let us start with $f(x) = 5x(1 - x)$, $x \in \mathbb{R}$. The graph of f is just a parabola looking down: Note that $f(x) = 5x(1 - x) < 5x$ when $x < 0$, so $f^n(x) < 5^n x$ in this

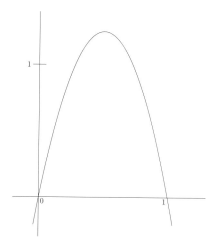

FIGURE 10. The graph of $f(x) = 5x(1 - x)$.

case, and, therefore, $f^n(x) \to -\infty$. If $x > 1$, then $f(x) < 0$ and, by the previous remark, we still have $f^n(x) \to -\infty$. Thus, the points x for which $f^n(x)$ does not tend to $-\infty$ must stay in the interval $[0, 1]$ after arbitrarily many iterations. Such points do exist. For example, the fixed point 0 of f satisfies this property, and so do all its pre-images.

To find the set of points with bounded iterates exactly, note that each such point should lie in any iterated pre-image $f^{-n}([0, 1])$. Moreover, since $f^{-1}([0, 1]) \subsetneq [0, 1]$, these pre-images form a nested sequence of closed sets. They can be easily seen on the graph. The set $f^{-1}([0, 1])$ is the union of 2 intervals $[0, x_1]$ and $[x_2, 1]$: The second pre-image $f^{-2}([0, 1])$ is a union of four intervals: and so on. The final

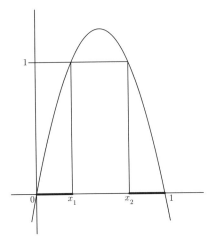

FIGURE 11. The set $f^{-1}([0,1])$.

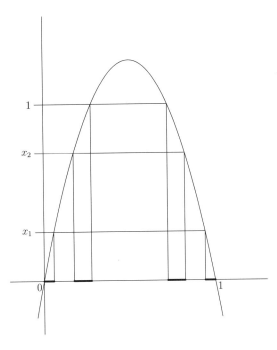

FIGURE 12. The set $f^{-2}([0,1])$.

intersection $\bigcap\limits_{n=1}^{\infty} f^{-n}([0,1])$ has pretty much the same structure as the middle third Cantor set.

The pictures become even more interesting if we consider the quadratic mapping on the complex plane. Let $f_c(z) = z^2 + c$, $(c \in \mathbb{C})$. The set of points $z \in \mathbb{C}$ such that the sequence z, $f_c(z)$, $f_c^2(z), \ldots$ does not tend to ∞ is called the *filled Julia set* of the mapping f_c. Several filled Julia sets corresponding to different values of c are shown below.

FIGURE 13. Julia sets.

Once we have a parametric set $\{f_c\}_c$ of mappings, we can also consider the set of the values of the parameter c for which f_c has a certain property. The famous Mandelbrot fractal on Figure 14 is defined as the set of all $c \in \mathbb{C}$ for which 0 belongs to the filled Julia set of f_c or, equivalently, for which the Julia set of f_c is connected.

One more example where infinite iteration arises naturally is the Newton method for finding roots of functions.

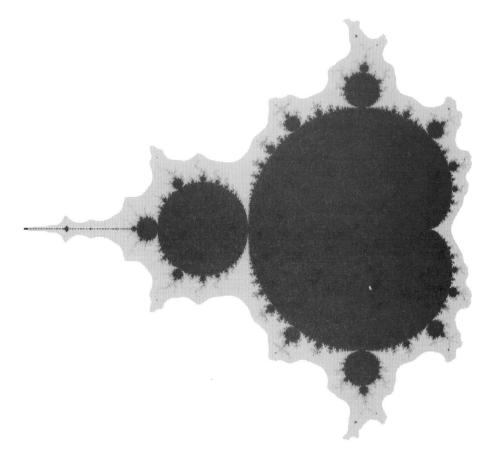

FIGURE 14. The Mandelbrot set

Suppose that we need to solve the equation $f(x) = 0$ and know some initial approximation x_0 to the root. Linearizing f at x_0, we get the equation

$$f(x_0) + f'(x_0)(x - x_0) = 0,$$

whose solution is $x_1 = x_0 - \frac{f(x_0)}{f'(x_0)}$. If the function f is nice enough, and x_0 is close enough to the root we are seeking, x_1 is closer to the root than x_0: However, that is not always the case:

The classical *Newton method* uses this iteration process to approximate the root. Note that, when the function f has several roots, it is by no means trivial to determine which of them the Newton method will converge to starting with some fixed x_0 (if it converges at all) and the set of initial values leading to a particular root is referred to as the *basin of attraction* of that root. This becomes an even more complicated question when we apply the Newton method to an analytic mapping on the complex plane. The basin of attraction of the root $z = 1$ of the function $f(z) = z^3 - 1$ is depicted in Figure 17. The boundary of this basin is very far from being smooth. It, too, has fractal structure.

For more information regarding fractals in the complex plane and their relation to dynamics, the reader is referred to [**B**] and [**Mi**].

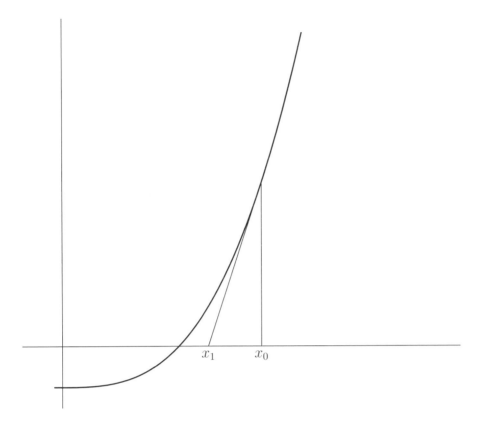

FIGURE 15. In a nice situation x_1 is closer to the root than x_0.

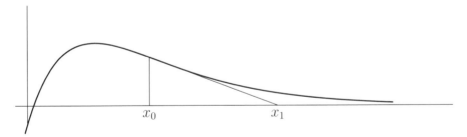

FIGURE 16. Bad situation.

FIGURE 17. The basin of attraction of the root $z = 1$.

CHAPTER 2

Dimension

Since the notion of a tangent plane is absent in the fractal setting, we need an alternative way to explain why a curve has dimension 1 and a surface has dimension 2. We will now discuss a construction that works in any compact metric space X.

Let A be a closed subset of X. Fix some radius $r > 0$. Let $N_r(A)$ be the least number of balls of radius r needed to cover A. If A is a rectifiable curve in $\mathbb{R}^n$, it is not hard to show that $N_r(A)$ is of order r^{-1}. If A is a surface, $N_r(A)$ is approximately r^{-2}. This suggests the idea to define the dimension of an arbitrary set as the number d for which $N_r(A) \approx r^{-d}$.

DEFINITION. *The limit $\lim\limits_{r \to 0^+} \frac{\log N_r(A)}{\log 1/r}$, when it exists, is called the Minkowski dimension of A and is denoted by M-$dim A$.*

Even if the limit fails to exist, we still can talk about the upper and the lower limits, which give us the upper and the lower Minkowski dimensions

$$\overline{M}\text{-}dim A = \limsup_{r \to 0^+} \frac{\log N_r(A)}{\log \frac{1}{r}} \qquad \text{and} \qquad \underline{M}\text{-}dim A = \liminf_{r \to 0^+} \frac{\log N_r(A)}{\log \frac{1}{r}}$$

correspondingly.

Since $N_r(A)$ is a non-increasing function of r and since $\log \frac{1}{r}$ changes very slowly, instead of the (upper or lower) limit along all values of r we can take the (upper or lower) limit along any sequence r_j such that $\frac{\log \frac{1}{r_{j+1}}}{\log \frac{1}{r_j}} \to 1$ as $j \to \infty$.

When $X = Q$ is the unit cube in $\mathbb{R}^n$, we can use cubic boxes instead of balls. More precisely, take some integer $p > 0$ and split Q into p^n equal subcubes with sidelength $\frac{1}{p}$. For $A \subset Q$, let $H_p(A)$ be the number of subcubes that intersect A.

DEFINITION. *The box dimension of A is*

$$B\text{-}dim A = \lim_{p \to +\infty} \frac{\log H_p(A)}{\log p}.$$

Again, in the cases when the limit fails to exist, we may talk about the upper and the lower box dimensions instead.

Since each cube with sidelength $\frac{1}{p}$ is contained in a ball of radius $\frac{\sqrt{n}}{p}$, say, we conclude that $N_{\frac{\sqrt{n}}{p}}(A) \leq H_p(A)$. Since each ball of radius $\frac{1}{p}$ can intersect at most some finite number $C(n)$ cubes of size $\frac{1}{p}$, we have

$$H_p(A) \leq C(n) N_{\frac{1}{p}}(A).$$

Taking the logarithm, dividing by $\log p$, and passing to the limit as $p \to +\infty$, we see that the box dimension we just defined coincides with the Minkowski dimension and the same is true for the upper and lower versions.

The next notion of dimension we will discuss is more complicated but also more useful. It is called the *Hausdorff dimension*. We start with defining sets of Hausdorff α-measure 0.

DEFINITION. *Fix $\alpha > 0$. Let X be a metric space and let A be any subset of X. We say that A is of Hausdorff α-measure 0 ($\mathcal{H}^\alpha(A) = 0$) if for every $\varepsilon > 0$, there exists a countable family of balls $B_{r_i}(x_i)$ ($x_i \in X$, $r_i > 0$) such that $A \subset \bigcup_i B_{r_i}(x_i)$ and $\sum_i r_i^\alpha < \varepsilon$.*

Note that if $\beta > \alpha > 0$ and $\mathcal{H}^\alpha(A) = 0$, then $\mathcal{H}^\beta(A) = 0$ as well. Indeed, take $\varepsilon \in (0, 1)$ and consider a covering of A by balls $B_{r_i}(x_i)$ with $\sum_i r_i^\alpha < \varepsilon$. Since $\varepsilon < 1$, it follows that $r_i < 1$ for all i, so $r_i^\beta \leq r_i^\alpha$, whence $\sum_i r_i^\beta \leq \sum_i r_i^\alpha < \varepsilon$ as well.

Thus, for every $A \subset X$, the set $\{\alpha > 0 : \mathcal{H}^\alpha(A) = 0\}$ is either empty, or a ray from some $\alpha_0 \geq 0$ to $+\infty$. In the second case, this $\alpha_0 = \inf\{\alpha > 0 : \mathcal{H}^\alpha(A) = 0\}$ is called the Hausdorff dimension of A and denoted by $dim\, A$. If the aforementioned set of α is empty, it is natural to say that $dim\, A = \infty$. Obviously, if $A \subset B$, then $dim\, A \leq dim\, B$ (every covering of B is a covering of A as well).

Now suppose that the lower Minkowski dimension of A is less than β. It means that $\frac{\log N_r(A)}{\log 1/r} < \beta$ along some sequence of values of $r > 0$ tending to 0. For every r in this sequence, we have $N_r(A) < r^{-\beta}$, so for every $\beta' > \beta$, the set A can be covered by balls with the sum of β' powers of radii not exceeding $N_r(A)r^{\beta'} \leq r^{\beta'-\beta} \to 0$ as $r \to 0$.

Thus, for every $\beta > \underline{M}\text{-}dim\, A$ and every $\beta' > \beta$, we have $\mathcal{H}^{\beta'}(A) = 0$, so $dim\, A \leq \beta$. Since β is arbitrary here, we conclude that $dim\, A \leq \underline{M}\text{-}dim\, A$ for all A.

Since the Minkowski dimension of the unit cube $Q \subset \mathbb{R}^n$ equals n, we have

$$dim\, A \leq \underline{M}\text{-}dim\, A \leq M\text{-}dim\, Q = n$$

for every $A \subset Q$.

The following Lemma allows one to estimate the Hausdorff dimension of A from below.

LEMMA 1. *Assume that there exists a measure μ on X such that $\mu(A) > 0$, and $\mu(B_r(x)) \leq Kr^\alpha$ for all $x \in X$, $r > 0$. Then $dim\, A \geq \alpha$.*

PROOF. For every covering $A \subset \bigcup_i B_{r_i}(x_i)$, we have

$$\mu(A) \leq \sum_i \mu(B_{r_i}(x_i)) \leq K \sum_i r_i^\alpha,$$

so it is impossible to make $\sum_i r_i^\alpha$ less than $K^{-1}\mu(A)$. Thus, we cannot have $\mathcal{H}^\alpha(A) = 0$, which means that $dim\, A \geq \alpha$. $\square$

Now let us compute the Hausdorff dimension of the middle-third Cantor set C.

On one hand, C is a subset of C_n and C_n consists of 2^n intervals of length 3^{-n}. Thus,

$$\underline{M}\text{-}dim\, C \leq \lim_{n \to \infty} \frac{\log(2^n)}{\log(3^n)} = \frac{\log 2}{\log 3}.$$

On the other hand, we can use the alternative description of C as the set of all numbers of the form $\sum_{k=1}^{\infty} \frac{2\varepsilon_k}{3^k}$, $\varepsilon_k \in \{0,1\}$, (see [**KF**], page 52, Example 4), and consider the mapping $f : [0,1] \backslash \mathbb{Q} \to [0,1]$ defined by $f\left(\sum_{k=1}^{\infty} \frac{\varepsilon_k}{2^k} \right) = \sum_{k=1}^{\infty} \frac{2\varepsilon_k}{3^k}$. Defining the measure μ on $[0,1] \setminus \mathbb{Q}$ by $\mu(B) = m(f^{-1}(B))$ where m is the one-dimensional Lebesgue measure, we see that $\mu(C) = m([0,1]\backslash\mathbb{Q}) = 1$. However, if I is any interval of length $r \in [\frac{1}{3^{l+1}}, \frac{1}{3^l})$, then the values of ε_k, with $k \leq l$ in the representation $\sum_{k=1}^{\infty} \frac{2\varepsilon_k}{3^k}$ of a number in I are completely determined. Indeed, if $\varepsilon_k' = \varepsilon_k''$ for $k < k_0 \leq l$, and $\varepsilon_{k_0}'' > \varepsilon_{k_0}'$, then

$$\sum_{k=1}^{\infty} \frac{2\varepsilon_k''}{3^k} - \sum_{k=1}^{\infty} \frac{2\varepsilon_k'}{3^k} = \sum_{k=k_0}^{\infty} \frac{2(\varepsilon_k'' - \varepsilon_k')}{3^k} \geq \frac{2}{3^{k_0}} - \sum_{k>k_0} \frac{2}{3^k} = \frac{1}{3^{k_0}} > r,$$

so the corresponding points cannot lie in I simultaneously. However, all numbers $\sum_{k=1}^{\infty} \frac{\varepsilon_k}{2^k}$ with fixed $\varepsilon_1, \ldots, \varepsilon_l$ are contained in an interval of length 2^{-l}, so $\mu(I) = m(f^{-1}(I)) \leq 2^{-l} \leq 2r^{\frac{\log 2}{\log 3}}$. Thus, by Lemma 1, $dim\, C \geq \frac{\log 2}{\log 3}$.

A similar argument shows that the Hausdorff dimension of the Serpinski gasket S equals $\frac{\log 3}{\log 2}$.

Note that the converse statement to the statement of Lemma 1 also holds for compact sets A: if $dim\, A = \alpha > 0$, then for every $\alpha' < \alpha$, there exists a measure μ on X satisfying $\mu(A) > 0$ and $\mu(B_r(x)) \leq Cr^{\alpha'}$ for all $x \in X$, $r > 0$. This result is called Frostman's lemma (see [**Mat**], pages 112-114) and Chapter 5 of these notes. Thus, at least for compact sets, the Hausdorff dimension is closely related to measures of restricted growth supported on a set.

Finally, we will introduce the *star dimension*. The starting point will be the same as in the case of the box dimension: we will partition a cube into p^n subcubes and count the number of subcubes intersecting $A \subset [0,1]^n$. The crucial difference is that now the cube to be partitioned is *not* the full cube $Q = [0,1]^n$, but an arbitrary subcube Q' of Q.

The formal definition is as follows. For a set $A \subset Q$ define $H_p^*(A)$ to be the largest number of subcubes that can intersect A in any partition of any cube $Q' \subset Q$ into p^n equal subcubes. Clearly if $A \neq \varnothing$, we have $1 \leq H_p^*(A) \leq p^n$.

LEMMA 2. *The limit* $\lim_{p \to \infty} \frac{\log H_p^*(A)}{\log p}$ *exists and is a number between* 0 *and* n.

PROOF. Since $0 \leq \frac{\log H_p^*(A)}{\log p} \leq n$ for all p, it suffices to establish the existence of the limit.

Fix any $p_0 > 1$. Let $\alpha = \frac{\log H_{p_0}^*(A)}{\log p_0}$. Take a large p of the form $p = p_0^m$, $m \in \mathbb{N}$. Note that $H_{pq}^*(A) \leq H_p^*(A)H_q^*(A)$ because we can think of the partition of a cube into $(pq)^n$ subcubes as of a preliminary partition into p^n subcubes followed by the partition of each of those subcubes into q^n subsubcubes. At most $H_p^*(A)$ of the subcubes in the preliminary partition can intersect A, and in each of those that do, at most $H_q^*(A)$ subsubcubes can intersect A.

Thus, by induction, we get $H_p^*(A) \leq H_{p_0}^*(A)^m$, whence

$$\frac{\log H_p^*(A)}{\log p} \leq \frac{m \log H_{p_0}^*(A)}{m \log p_0} = \alpha.$$

Now consider $p \in [p_0^m, p_0^{m+1})$. Note that the small cubes in the partition of a cube Q' into p^n equal subcubes can be gathered into at most p_0^n (possibly overlapping) cubic blocks of $(p_0^m)^n$ cubes each. Hence, $H_p^*(A) \leq p_0^n H_{p_0}^*(A)^m$ in this case and

$$\frac{\log H_p^*(A)}{\log p} \leq \frac{m \log H_{p_0}^*(A)}{m \log p_0} + \frac{n \log p_0}{\log p}.$$

Thus,

$$\limsup_{p \to \infty} \frac{\log H_p^*(A)}{\log p} \leq \alpha + \limsup_{p \to \infty} \frac{n \log p_0}{\log p} = \alpha.$$

Since p_0 was arbitrary here, we conclude that

$$\limsup_{p \to \infty} \frac{\log H_p^*(A)}{\log p} \leq \inf_{p > 1} \frac{\log H_p^*(A)}{\log p} \leq \liminf_{p \to \infty} \frac{\log H_p^*(A)}{\log p}.$$

DEFINITION. *The limit* $\lim_{p \to \infty} \frac{\log H_p^*(A)}{\log p}$ *whose existence has been proved in Lemma 2 is called the star dimension of A and denoted by* $\dim^* A$.

EXAMPLE 1. *Let $n = 1$ and $A = \{0\} \cup \{\frac{1}{k} : k \in \mathbb{N}\} \subset [0,1]$.*

In this case $\dim A = 0$ (because A is countable). To find the $M\text{-}\dim A$, note that when we try to cover A by balls of some small radius $r > 0$, every point $\frac{1}{k}$ with $\frac{1}{k} - \frac{1}{k+1} = \frac{1}{k(k+1)} > 2r$ needs an individual ball, so $N_r(A) \geq cr^{-\frac{1}{2}}$ for small $r > 0$. On the other hand, the points $\frac{1}{k}$ with $\frac{1}{k(k+1)} < 2r$ are all contained in the interval $[0, C\sqrt{r}]$, which can be covered by $Cr^{-\frac{1}{2}}$ intervals of length $2r$. Thus, $N_r(A) \leq Cr^{-\frac{1}{2}}$. These estimates imply immediately that $M\text{-}\dim A = \frac{1}{2}$. Finally, note that if we take any integer $p > 1$ and partition the interval $[0, \frac{1}{p}]$ into p equal subintervals, each of them will contain a point in A, so $H_p^*(A) = p$ for all p and $\dim^* A = 1$.   □

We now formulate the first theorem to be proved by ergodic theoretic means. Although this theorem is not hard to prove by standard considerations, it will serve to illustrate the machinery that will play a crucial role in our later discussion.

THEOREM 1. *If A is a non-empty closed subset of $Q = [0,1]^n$, then $\dim^* A = \max\{\dim A' : A'$ is a micro-set of $A\}$.*

Note that, in particular, the theorem claims that the micro-set of maximal dimension does exist. Also, the possibility to pass to the limit in the Hausdorff metric is crucial here. As the last example shows, we cannot restrict ourselves to mini-sets instead of micro-sets: all mini-sets of $A = \{0\} \cup \{\frac{1}{k} : k \in \mathbb{N}\}$ are countable and, thereby, have Hausdorff dimension 0. Note however that in the Hausdorff metric, we have $\lim_{\lambda \to +\infty} (\lambda A \cap [0,1]) = [0,1]$, so the micro-sets of A include the whole interval $[0,1]$, which has Hausdorff dimension 1.

CHAPTER 3

Trees and Fractals

Everywhere below, the word *"fractal"* will mean a non-empty closed subset of $Q = [0,1]^n$. We start with introducing the notion of a tree.

Let Λ be any finite set (*alphabet*).

Let $\Lambda^* = \bigcup_{l=0}^{\infty} \Lambda^l$ be the set of all finite words in the alphabet Λ. The set Λ^* can be viewed as a semigroup with respect to the concatenation operation $u, v \mapsto uv$. The empty word $\varnothing$ is the unit of this semigroup.

DEFINITION. *A set $\tau \subset \Lambda^*$ is a Λ-tree if*

1) the empty word belongs to τ;

2) for every $u \in \tau$ there exists $x \in \Lambda$ such that $ux \in \tau$ (every word can be extended to a longer one);

3) if $uv \in \tau$, then $u \in \tau$ (the beginning of every word in a tree is also in the tree).

A Λ-tree can be visualized as a directed tree without leaves whose edges are oriented away from the root and whose vertices are labeled by elements of Λ so that the labels at the ends of the edges going out of each vertex are all different. Sometimes, when the choice of Λ is clear or irrelevant, we will write just "tree" instead of "Λ-tree".

The boundary $\partial\tau$ of a tree τ is the set of all infinite words $\omega = \xi_1\xi_2\xi_3\ldots$, $\xi_j \in \Lambda$, such that every beginning $w = \xi_1\xi_2\xi_3\ldots\xi_m$ belongs to τ. In the graphical representation above, $\partial\tau$ is just the set of words one can read when going over all possible infinite paths in the graph starting at the root $\varnothing$.

Our next aim will be to define the *dimension of a tree*.

DEFINITION. *The Minkowski dimension of a tree τ is the limit*

$$M\text{-}dim\,\tau = \lim_{l \to \infty} \frac{\log N_l(\tau)}{l},$$

where $N_l(\tau)$ is the number of words w in τ of length $l(w) = l$.

As usual, if the limit fails to exist, we can talk about the upper and lower Minkowski dimension instead.

To define the Hausdorff dimension, we need the notion of a *section*.

DEFINITION. *Let τ be a tree. A finite subset $S \subset \tau$ is called a section if every sufficiently long word $w \in \tau$ has an initial segment in S, i.e., if there exists L such that every word $w \in \tau$ of length $l(w) \geq L$ can be written as uv with $u \in S$. A section S is called minimal if no proper subset of S is a section.*

Figure 2 illustrates this definition:

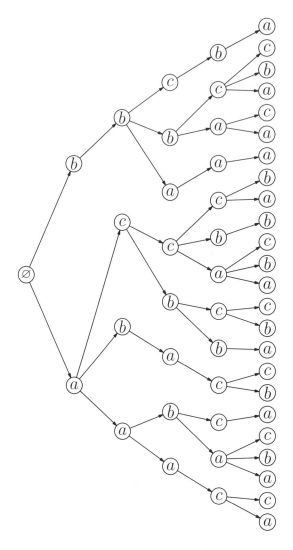

FIGURE 1. A Λ-tree over the alphabet $\{a, b, c\}$.

DEFINITION. *The Hausdorff dimension of a tree τ is the infimum of all $\lambda > 0$ such that for every $\varepsilon > 0$, there exists a section S of τ with $\sum\limits_{w \in S} e^{-\lambda l(w)} < \varepsilon$.*

Clearly, we can restrict ourselves to minimal sections in this definition.

Note that for every $l \geq 0$, the set S_l of all words of length l in a tree is a section, (we shall call any such section a *flat section*).

If $\lambda' > \lambda > \underline{M}\text{-}dim\,\tau$, we can find arbitrarily large l such that $N_l(\tau) \leq e^{\lambda l}$. For such l, we have

$$\sum_{w \in S_l} e^{-\lambda' l(w)} \leq e^{\lambda l} e^{-\lambda' l} = e^{-(\lambda' - \lambda)l} \to 0 \quad \text{as} \quad l \to \infty.$$

Thus, we always have $dim\,\tau \leq \underline{M}\text{-}dim\,\tau$.

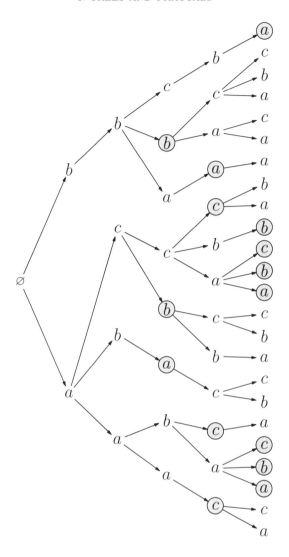

FIGURE 2. A minimal section of a tree.

Fix an integer $p > 1$. Let $\Lambda = \{0, 1, \ldots, p - 1\}$. With every finite word $w = \xi_1 \ldots \xi_l \in \Lambda^*$ of length $l \geq 1$, we associate the closed interval

$$I_w = \left[\frac{\xi_1}{p} + \cdots + \frac{\xi_{l-1}}{p^{l-1}} + \frac{\xi_l}{p^l}, \frac{\xi_1}{p} + \cdots + \frac{\xi_{l-1}}{p^{l-1}} + \frac{\xi_l + 1}{p^l} \right]$$

of length p^{-l}. We also put $I_\varnothing = [0, 1]$.

Let $A \subset [0, 1]$ be a closed non-empty set.

CLAIM 1. *Let $\widehat{\tau}_A$ be the set of all words w such that $I_w \cap A \neq \varnothing$. Then $\widehat{\tau}_A$ is a tree.*

PROOF. Since $I_\varnothing = [0, 1]$ and $A \neq \varnothing$, we have $\varnothing \in \widehat{\tau}_A$. Since for every $w \in \Lambda^*$, we have $I_w = \bigcup_{x \in \Lambda} I_{wx}$, we may have $I_w \cap A \neq \varnothing$ only if $I_{wx} \cap A \neq \varnothing$ for some

$x \in \Lambda$. Finally, since $I_{uv} \subset I_u$ for all $u, v \in \Lambda^*$, the condition $I_{uv} \cap A \neq \varnothing$ implies $I_u \cap A \neq \varnothing$. □

Alternatively, the tree $\widehat{\tau}_A$ associated with a closed set $A \subset [0, 1]$ can be described as the set of all initial segments of all possible p-ary expansions $t = \sum_{k=1}^{\infty} \frac{\xi_k}{p^k}$ of all points $t \in A$.

Our next goal will be to relate the dimension of A to the dimension of the corresponding tree $\widehat{\tau}_A$.

We shall start with the Minkowski dimensions. Note that for every $l \geq 1$, the set A is covered by the intervals I_w with $w \in S_l = \{u \in \widehat{\tau}_A : l(u) = l\}$. All those intervals have length p^{-l}, so $N_{p^{-l}}(A) \leq N_l(\widehat{\tau}_A)$, whence

$$\frac{\log N_{p^{-l}}(A)}{\log(p^l)} \leq \frac{1}{\log p} \frac{\log N_l(\widehat{\tau}_A)}{l}.$$

Passing to the (upper or lower) limit as $l \to \infty$, we see that

$$\overline{M}\text{-}dim\, A \leq \frac{1}{\log p} \overline{M}\text{-}dim\, \widehat{\tau}_A \quad \text{and} \quad \underline{M}\text{-}dim\, A \leq \frac{1}{\log p} \underline{M}\text{-}dim\, \widehat{\tau}_A.$$

On the other hand, every one-dimensional ball of radius p^{-l} (i.e., an open interval of length $2p^{-l}$) can intersect at most 3 intervals I_w with $l(w) = l$. Thus, we have $N_l(\widehat{\tau}_A) \leq 3N_{p^{-l}}(A)$ and

$$\frac{\log N_l(\widehat{\tau}_A)}{l} \leq \log p \left[\frac{\log N_{p^{-l}}(A)}{\log(p^l)} + \frac{\log 3}{\log(p^l)} \right].$$

Passing to the (upper or lower) limit as $l \to \infty$, we conclude that

$$\overline{M}\text{-}dim\, \widehat{\tau}_A \leq (\overline{M}\text{-}dim\, A) \log p \quad \text{and} \quad \underline{M}\text{-}dim\, \widehat{\tau}_A \leq (\underline{M}\text{-}dim\, A) \log p.$$

Juxtaposing these inequalities with the ones we obtained earlier, we conclude that

$$\overline{M}\text{-}dim\, \widehat{\tau}_A = (\overline{M}\text{-}dim\, A) \log p, \quad \text{and} \quad \underline{M}\text{-}dim\, \widehat{\tau}_A = (\underline{M}\text{-}dim\, A) \log p.$$

The same result holds for the Hausdorff dimensions though the proof is slightly more complicated.

LEMMA 3. *For every non-empty closed set $A \subset [0, 1]$, we have*

$$dim\, \widehat{\tau}_A = (dim\, A) \log p.$$

PROOF. We start with noting that for every section S of a tree $\widehat{\tau}_A$, the intervals I_w, $w \in S$, cover A. Indeed, take any point $t \in A$. By the definition of $\widehat{\tau}_A$, for every L, one can find a word $u \in \widehat{\tau}_A$ with $l(u) > L$ such that $t \in I_u$. However, if L is large enough, such a word u can be written as $u = wv$ with $w \in S$. Since $I_u \subset I_w$, we conclude that $t \in I_w$.

Now, assume that $\lambda > dim\, \widehat{\tau}_A$. Then, for every $\varepsilon > 0$, we can find a section S of τ with $\sum_{w \in S} e^{-\lambda l(w)} < \varepsilon$. The corresponding covering of A by the intervals I_w, $w \in S$, satisfies

$$\sum_{w \in S} (diam\, I_w)^{\frac{\lambda}{\log p}} = \sum_{w \in S} e^{-\lambda l(w)} < \varepsilon.$$

Thus, $\mathcal{H}^{\frac{\lambda}{\log p}}(A) = 0$ and we conclude that $dim\, A \leq \frac{dim\, \widehat{\tau}_A}{\log p}$.

Assume now that $\lambda > dim\, A$. Then, for every $\epsilon > 0$, we can find a covering of A by balls $B(x_j, r_j)$ (which in the one-dimensional situation we are considering are just the intervals $(x_j - r_j, x_j + r_j)$) so that $\sum_j r_j^\lambda < \varepsilon$.

Since A is compact we can assume that this covering is finite. Each ball $B(x_j, r_j)$ intersects at most 3 intervals I_w with $w \in \widehat{\tau}_A$ and $p^{-l(w)} \in [r_j, pr_j)$. Let S be the set of all words $w \in \widehat{\tau}_A$ that can arise in this way. We clearly have

$$\sum_{w \in S} e^{-\lambda \log p\, l(w)} = \sum_{w \in S} p^{-\lambda l(w)} \leq 3p^\lambda \sum_j r_j^\lambda,$$

so to prove the inequality $dim\, \widehat{\tau}_A \leq (dim\, A) \log p$, it suffices to show that S is a section.

To this end, take any $u \in \widehat{\tau}_A$ such that $p^{-l(u)} < \min_j r_j$. Let t be any point in $I_u \cap A$. Then there exists j such that $t \in B(x_j, r_j)$. Since $p^{-l(u)} < r_j$, the word u can be written as $u = wv$ with $p^{-l(w)} \in [r_j, pr_j)$. Since $\widehat{\tau}_A$ is a tree, we have $w \in \widehat{\tau}_A$. Note now that $I_w \supset I_u \ni t$, so $I_w \cap B(x_j, r_j) \neq \varnothing$. Thus, $w \in S$. $\qquad\square$

CHAPTER 4

Invariant Sets

Fix an integer $p > 1$ and consider the mapping $T_p : [0,1] \to [0,1]$ defined by $T_p x = \{px\}$ where for real t, $\{t\}$ denotes the fractional part of $t : 0 \leq \{t\} < 1$, and $t - \{t\}$ an integer. Suppose that A is a closed subset of $[0,1]$ satisfying $T_p A = A$.

PROPOSITION 1. $\dim A = M\text{-}\dim A$.

PROOF. We always have $\dim A \leq \underline{M}\text{-}\dim A$. Thus, it suffices to prove that $\overline{M}\text{-}\dim A \leq \dim A$. Since $\overline{M}\text{-}\dim A = \frac{1}{\log p}\overline{M}\text{-}\dim \widehat{\tau}_A$ and $\dim A = \frac{1}{\log p}\dim \widehat{\tau}_A$, we can check this inequality for $\widehat{\tau}_A$ instead of A. Note that if we have a word $w = \xi_1 \xi_2 \ldots \xi_m \in \widehat{\tau}_A$, then the invariance of A under T_p implies that $w' = \xi_2 \ldots \xi_m \in \widehat{\tau}_A$ (if w is an initial segment in the p-ary decomposition of $t \in A$, then w' is an initial segment in the p-ary decomposition of $T_p t$).

It follows by induction that if $uv \in \widehat{\tau}_A$, then $v \in \widehat{\tau}_A$ as well. Combined with the defining properties of a tree, this means that every subword of every word in $\widehat{\tau}_A$ is again in $\widehat{\tau}_A$.

Assume now that $\dim \widehat{\tau}_A < \lambda$. Then, for every $\varepsilon > 0$, there exists a section S of $\widehat{\tau}_A$ such that $\sum_{w \in S} e^{-\lambda l(w)} < \varepsilon$. If $\varepsilon < 1$, this section cannot contain the empty word, so

$$0 < \min_{w \in S} l(w) \leq \max_{w \in S} l(w) = L < \infty.$$

Now take a large integer l and consider the set $\widetilde{S}$ of all words $w \in \widehat{\tau}_A$ that can be written as $w = v_1 v_2 \ldots v_k$, where $v_j \in S$, $l(v_1 \ldots v_k) \geq l$ and $l(v_1 \ldots v_{k-1}) < l$. Note that we have $l \leq l(w) \leq l + L$ for every $w \in \widetilde{S}$.

We want to show that $\widetilde{S}$ is a section of $\widehat{\tau}_A$.

Take any word $u \in \widehat{\tau}_A$ of length $l(u) \geq l + L$. Since S is a section and $l(u) \geq L$, we can write u as $v_1 u_1$ where $v_1 \in S$ (formally, we can only guarantee such representation if $l(u) > L_0$ with some L_0 but if there exists a word u_0 of length L that has no initial segment in S, then any continuation of u_0, which can be arbitrarily long, has no initial segment in S either).

If $l(v_1) < l$, then $l(u_1) \geq L$ and, since $u_1 \in \widehat{\tau}_A$, we can apply the same argument to u_1 and get a representation $w = v_1 v_2 u_2$ with $v_2 \in S$. We can continue this way as long as $l(v_1 \ldots v_k) < l$. Note that, since $\varnothing \notin S$, this condition cannot hold forever, so at some step we will have the inequality $l(v_1 \ldots v_k) \geq l$ for the first time. But then $v_1 \ldots v_k \in \widetilde{S}$ and $v_1 \ldots v_k$ is an initial segment of u.

The next observation is that for every $k \geq 1$, we have

$$\sum_{v_j \in S,\, j=1,\ldots,k} e^{-\lambda l(v_1 \ldots v_k)} = \sum_{v_j \in S,\, j=1,\ldots,k} e^{-\lambda \sum_{j=1}^{k} l(v_j)} = \left[\sum_{v \in S} e^{-\lambda l(v)}\right]^k < \varepsilon^k.$$

Summing over k, we obtain the inequality

$$\sum_{v \in \widetilde{S}} e^{-\lambda l(v)} \le \varepsilon + \varepsilon^2 + \cdots = \frac{\varepsilon}{1-\varepsilon}.$$

Now consider any word $u \in \widehat{\tau}_A$ of length $l(u) = l + L$. Since $\widetilde{S}$ is a section, u can be written as $u = wv$ with $w \in \widetilde{S}$. Note that $l(v) = l(u) - l(w) \le l + L - l = L$, so we can have only $1 + p + p^2 + \cdots + p^L = \frac{p^{L+1}-1}{p-1}$ different possibilities for v in such representations. Thus,

$$e^{-\lambda(l+L)} N_{l+L}(\widehat{\tau}_A) = \sum_{u \in \widehat{\tau}_A, \, l(u) = l+L} e^{-\lambda l(u)} \le \frac{p^{L+1}-1}{p-1} \sum_{w \in \widetilde{S}} e^{-\lambda l(w)} \le \frac{p^{L+1}-1}{p-1} \frac{\varepsilon}{1-\varepsilon},$$

whence

$$\frac{\log N_{l+L}(\widehat{\tau}_A)}{l+L} \le \lambda + \frac{C}{l+L}.$$

Taking the upper limit as $l \to \infty$, we obtain $\overline{M}\text{-}dim\,\widehat{\tau}_A \le \lambda$. Since $\lambda > dim\,\widehat{\tau}_A$ was arbitrary, we conclude that $\overline{M}\text{-}dim\,\widehat{\tau}_A \le dim\,\widehat{\tau}_A$ as desired. □

CHAPTER 5

Probability Trees

DEFINITION. *A probability tree on the set* Λ^* *of all finite words in the alphabet* Λ *is any function* $\Theta : \Lambda^* \to [0,1]$ *such that*
1) $\Theta(\varnothing) = 1$;
2) $\Theta(w) = \sum\limits_{x \in \Lambda} \Theta(wx)$ *for every* $w \in \Lambda^*$.

If Θ is a probability tree, then $\tau = \{w \in \Lambda^* : \Theta(w) > 0\}$ is a tree. Our first observation is that if S is a section of τ, then $\sum\limits_{w \in S} \Theta(w)$ cannot be small.

CLAIM 2. *For every section* S *of* τ, *we have* $\sum\limits_{w \in S} \Theta(w) \geq 1$. *Moreover, if* S *is minimal, then* $\sum\limits_{w \in S} \Theta(w) = 1$.

PROOF. Let $L = \max\limits_{w \in S} l(w)$. Property 2) of a probability tree implies by induction that for every $k \geq 1$, and every $w \in \Lambda^*$, we have

$$\Theta(w) = \sum_{v \in \Lambda^*, \, l(v)=k} \Theta(wv) = \sum_{v \in \Lambda^*, \, l(v)=k, \, wv \in \tau} \Theta(wv).$$

Now consider all $u \in \Lambda^*$ with $l(u) = L$. On one hand,

$$1 = \Theta(\varnothing) = \sum_{u \in \Lambda^* : l(u)=L} \Theta(u) = \sum_{u \in \tau : l(u)=L} \Theta(u).$$

On the other hand, each $u \in \tau$ with $l(u) = L$ can be written as $u = wv$ with some $w \in S$, $v \in \Lambda^*$. Moreover, if w is fixed, the length of v is also determined by $l(v) = L - l(w)$. Thus,

$$\sum_{u \in \tau : l(u)=L} \Theta(u) \leq \sum_{w \in S} \sum_{v \in \Lambda^* : l(v)=L-l(w)} \Theta(wv) = \sum_{w \in S} \Theta(w)$$

and we conclude that $\sum\limits_{w \in S} \Theta(w) \geq 1$.

If S is a minimal section, then the representation $u = wv$ ($w \in S$, $v \in \Lambda^*$) is unique, so this inequality becomes an identity. $\qquad \square$

We can relate the Hausdorff dimension of τ to the exponential rate of decay of Θ in the following way.

CLAIM 3. *If* $\Theta(w) \leq Ce^{-\lambda l(w)}$ *for all* $w \in \Lambda^*$, *then* $\dim \tau \geq \lambda$.

PROOF. If $\Theta(w) \leq Ce^{-\lambda l(w)}$ for every $w \in \Lambda^*$, it follows that every section S of τ satisfies $\sum\limits_{w \in S} e^{-\lambda l(w)} \geq \frac{1}{C}$, which immediately implies that $\dim \tau \geq \lambda$. $\qquad \square$

The following propositions shows that the converse statement is also true.

PROPOSITION 2. *For any tree τ with $\dim \tau > \lambda$, we can find a probability tree Θ such that $\Theta(w) \leq C e^{-\lambda l(w)}$ for all $w \in \Lambda^*$ and $\tau \supset \{w \in \Lambda^* : \Theta(w) > 0\}$.*

PROOF. Since $\dim \tau > \lambda$, there exists $\varepsilon > 0$ such that for every section S of τ, we have $\sum_{w \in S} e^{-\lambda l(w)} \geq \varepsilon$. Put $C = \frac{1}{\varepsilon}$. Note, first of all, that it suffices to construct a subadditive function $\widetilde{\Theta} : \Lambda^* \to [0,1]$ (i.e., a function satisfying $\widetilde{\Theta}(w) \leq \sum_{x \in \Lambda} \widetilde{\Theta}(wx)$ for all $w \in \Lambda^*$) such that $\widetilde{\Theta}(\varnothing) = 1$, $\widetilde{\Theta}(w) \leq C e^{-\lambda l(w)}$ for all $w \in \Lambda^*$, and $\{w \in \Lambda^* : \widetilde{\Theta}(w) > 0\} \subset \tau$.

Indeed, let $\widetilde{\Theta}$ be a subadditive function with the above properties. For each $w \in \Lambda^*$, define the subadditivity factor $\gamma(w) \in [0,1]$ by $\widetilde{\Theta}(w) = \gamma(w) \sum_{x \in \Lambda} \widetilde{\Theta}(wx)$ and the cumulative subadditivity factor $\Gamma(w) \in [0,1]$ by $\Gamma(w) = \prod_{w'} \gamma(w')$ where the product is taken over all initial segments w' of w excluding w itself (we follow the usual convention that the empty product equals 1, so $\Gamma(\varnothing) = 1$).

Now put $\Theta(w) = \Gamma(w)\widetilde{\Theta}(w)$. Clearly, $\Theta(w) \leq \widetilde{\Theta}(w)$, so $\{w : \Theta(w) > 0\} \subset \tau$ and $\Theta(w) \leq C e^{-\lambda l(w)}$. Also $\Theta(\varnothing) = \Gamma(\varnothing)\widetilde{\Theta}(\varnothing) = 1 \cdot 1 = 1$. Finally,

$$\sum_{x \in \Lambda} \Theta(wx) = \sum_{x \in \Lambda} \Gamma(wx)\widetilde{\Theta}(wx) = \sum_{x \in \Lambda} \Gamma(w)\gamma(w)\widetilde{\Theta}(wx)$$

$$= \Gamma(w)\gamma(w)\sum_{x \in \Lambda} \widetilde{\Theta}(wx) = \Gamma(w)\widetilde{\Theta}(w) = \Theta(w),$$

so Θ is a probability tree.

To obtain $\widetilde{\Theta}$, we will first construct a subadditive function $\widetilde{\Theta}_N$ on $\bigcup_{n \leq N} \Lambda^n$.

We start with putting

$$\widetilde{\Theta}_N(w) = \begin{cases} C e^{-\lambda N}, & w \in \Lambda^N \cap \tau \\ 0, & w \in \Lambda^N \setminus \tau \end{cases}$$

and proceed backwards by induction: if $\widetilde{\Theta}_N$ is already defined on Λ^{n+1}, we define it on Λ^n by $\widetilde{\Theta}_N(w) = \min(C e^{-\lambda n}, \sum_{x \in \Lambda} \widetilde{\Theta}_N(wx))$.

Clearly Θ_N is subadditive and $\widetilde{\Theta}_N \geq \widetilde{\Theta}_{N+1}$ on $\bigcup_{n \leq N} \Lambda^n$. Thus, for each $w \in \Lambda^*$, $\widetilde{\Theta}_N(w)$ is well-defined as soon as $N \geq l(w)$ and is non-increasing as a function of N. So, $\widetilde{\Theta} = \lim_{N \to \infty} \widetilde{\Theta}_N$ exists, is subadditive, and satisfies $\widetilde{\Theta}(w) \leq C e^{-\lambda l(w)}$. The only problem is that the equality $\widetilde{\Theta}(\varnothing) = 1$ may fail. If $\widetilde{\Theta}(\varnothing) > 1$, it is not a serious obstacle because then we can redefine $\widetilde{\Theta}(\varnothing)$ to 1 preserving subadditivity. However, if we have $\widetilde{\Theta}(\varnothing) < 1$, it is a real problem. We will rule this case out by showing that $\widetilde{\Theta}_N(\varnothing) \geq 1$ for all N.

Let γ_N be the subadditivity factor for $\widetilde{\Theta}_N$, i.e.,

$$\gamma_N : \bigcup_{n \leq N-1} \Lambda^n \to [0,1], \qquad \widetilde{\Theta}_N(w) = \gamma_N(w) \sum_{x \in \Lambda} \widetilde{\Theta}_N(wx)$$

for all $w \in \bigcup_{n \leq N-1} \Lambda^n$.

Define S_N as the set of all words w in $\tau \cap \left(\bigcup_{n \leq N} \Lambda^n \right)$ such that $\gamma_N(w') = 1$ for all initial segments w' of w excluding w itself and either $\gamma_N(w) < 1$, or $l(w) = N$. Then S_N is a *minimal* section of τ and $\widetilde{\Theta}_N(w) = Ce^{-\lambda l(w)}$ for every $w \in S_N$.

Since $C \sum_{w \in S} e^{-\lambda l(w)} \geq 1$ for all sections S of τ (see Claim 2), we will be done if we show that $\widetilde{\Theta}_N(\varnothing) = \sum_{w \in S_N} \widetilde{\Theta}_N(w)$.

To this end, note that, since S_N is a minimal section of τ, each word $w \in \tau$ either has a proper initial segment in S_N, or can be extended (possibly trivially) to a word in S_N, but not both.

Let S_N^* be the set of the words in τ with the latter property. For each $w \in S_N^*$, denote by S_N^w the set of all extensions of w that lie in S_N. Observe that if $w \in S_N^* \setminus S_N$, then $wx \in S_N^*$ for all $x \in \Lambda$ such that $wx \in \tau$. Moreover, in this case we have $\gamma_N(w) = 1$ and

$$(1) \qquad \widetilde{\Theta}_N(w) = \sum_{x \in \Lambda} \widetilde{\Theta}_N(wx) = \sum_{x \in \Lambda, \, wx \in \tau} \widetilde{\Theta}_N(wx)$$

(because, clearly, $\widetilde{\Theta}_N$ is 0 on $\left(\bigcup_{n \leq N} \Lambda^n \right) \setminus \tau$).

The last property allows us to prove by backward induction in $l(w)$ that for every $w \in S_N^*$, we have $\widetilde{\Theta}_N(w) = \sum_{u \in S_N^w} \widetilde{\Theta}_N(u)$.

Indeed, if $l(w) = N$, then the only extension of w contained in S_N that can possibly be there is w itself, so the statement is trivial in this case.

Suppose that the statement is true for all $w \in S_N^*$ with $l(w) = m + 1$. Take $w \in S_N^*$ with $l(w) = m$. If $w \in S_N$, then $S_N^w = \{w\}$ (the existence of any non-trivial extensions would imply that the section S_N is not minimal), so in this case there is nothing to prove. Otherwise, $w \in S_N^* \setminus S_N$ and we can use (1) and apply the induction assumption to each word wx for which $wx \in \tau$ to get the desired inequality.

Finally, it remains to note that we always have $\varnothing \in S_N^*$ and $S_N^\varnothing = S_N$. $\qquad \square$

The result we have just proved is an exact analogue of the Frostman lemma for trees. Moreover, we can formally derive it from the Frostman lemma if we consider the boundary $\partial \tau$ as a metric space with the distance between two infinite words $\omega' = \xi_1' \xi_2' \ldots$ and $\omega'' = \xi_1'' \xi_2'' \ldots$ equal to e^{-k} where k is the minimal index for which $\xi_k' \neq \xi_k''$.

Conversely, introducing a "tree structure" on an arbitrary compact subset of $\mathbb{R}^n$ (similarly to our passage from A to $\widehat{\tau}_A$), we can derive the Frostman lemma for sets from our tree version. We leave both these derivations to the reader as exercises.

We will now make a useful observation about the dimensions of Cartesian products.

Let X, Y be two metric spaces. Define the distance d in $X \times Y$ by

$$d((x_1, y_1), (x_2, y_2)) = \max(d_X(x_1, x_2), d_Y(y_1, y_2)).$$

Let $A \subset X$, $B \subset Y$. Note that if $A \subset \bigcup_i B_r(x_i)$ and $B \subset \bigcup_j B_r(y_j)$, then $A \times B \subset \bigcup_{i,j} B_r((x_i, y_j))$, so $N_r(A \times B) \leq N_r(A)N_r(B)$ for all $r > 0$. Thus, we

have
$$\overline{M}\text{-}dim\,(A \times B) \leq \overline{M}\text{-}dim\,A \,+\, \overline{M}\text{-}dim\,B.$$

On the other hand, assume that μ and ν are finite measures on X and Y such that $\mu(B_r(x)) \leq r^\alpha$ for all $x \in X$, $r > 0$ and $\nu(B_r(y)) \leq r^\beta$ for all $y \in Y$, $r > 0$, but $\mu(A)$, $\nu(B) > 0$. Then the product measure $\mu \times \nu$ satisfies

$$(\mu \times \nu)(B_r((x,y))) = \mu(B_r(x))\nu(B_r(y)) \leq r^{\alpha+\beta}$$

and $(\mu \times \nu)(A \times B) = \mu(A)\nu(B) > 0$, so we conclude that

$$dim\,(A \times B) \geq dim\,A + dim\,B.$$

CHAPTER 6

Galleries

DEFINITION. *A family Γ of fractals is called a gallery if it is closed in the Hausdorff metric and for every $A \in \Gamma$, all mini-sets of A are also in Γ.*

Note that since micro-sets are defined as limits of mini-sets, with any fractal A, the gallery Γ must contain all its micro-sets as well. As a matter of fact, if A is any fractal, the least gallery Γ containing A is the one consisting of all micro-sets of A.

Since the set of micro-sets of A is just the closure of the set of all mini-sets of A, it is closed. Thus, the only thing one needs to prove here is the following lemma.

LEMMA 4. *A mini-set of a micro-set of A is a micro-set of A.*

PROOF. The difficulty is that the convergence of A_n to B in the Hausdorff metric does not necessarily imply the convergence of $A_n \cap Q$ to $B \cap Q$ (see Figure 1). So we have to be rather careful when constructing a sequence of mini-sets of A converging to a mini-set of a micro-set of A.

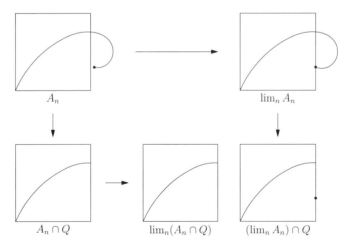

FIGURE 1. The set $(\lim_n A_n) \cap Q$ can have extra points compared to the set $\lim_n (A_n \cap Q)$.

Let B be a micro-set of A. Let $Q' \subset Q$ be the cube defining a mini-set B' of B, i.e., $B' = (TB) \cap Q = T(B \cap Q')$ where T is an affine mapping of the form $Tx = \lambda x + u$ such that $TQ' = Q$.

For $\varepsilon > 0$, put

$$Q'_\varepsilon = (1 - \varepsilon)Q' + \varepsilon Q = \{(1 - \varepsilon)x' + \varepsilon x : x' \in Q', x \in Q\}.$$

The cube Q'_ε contains Q' and is contained in Q. Also, $Q'_{\varepsilon_1} \subset Q'_{\varepsilon_2}$ for $\varepsilon_1 \le \varepsilon_2$ and $\bigcap_\varepsilon Q'_\varepsilon = Q'$.

FIGURE 2. The cube Q'_ϵ for various locations of the cube Q'.

Note first of all that $B \cap Q'_\varepsilon \to B \cap Q'$ as $\varepsilon \to 0$. Indeed, we obviously have $B \cap Q'_\varepsilon \supset B \cap Q'$. Assume now that for some sequence $\varepsilon_j \to 0+$, there exist points $x_j \in B \cap Q'_{\varepsilon_j}$ such that $d(x_j, B \cap Q') > \delta$ with some fixed $\delta > 0$. Since $x_j \in Q$ and Q is compact, we can assume without loss of generality that $x_j \to x$ as $j \to \infty$. However, since B is closed and Q'_{ε_j} shrink to Q', we must have $x \in B \cap Q'$. Then, for sufficiently large j, we have $d(x_j, B \cap Q) \le d(x_j, x) < \delta$, which is impossible.

Now let A_n be a sequence of subsets of Q converging to B. We claim that for every $\delta > 0$, we can choose $\varepsilon > 0$ and n such that $A_n \cap Q'_\varepsilon$ is δ-close to $B \cap Q'$.

To show that, start with choosing $\varepsilon > 0$ so small that $B \cap Q'_{2\varepsilon}$ is contained in the $\frac{\delta}{3}$-neighborhood of $B \cap Q'$.

There exists $\rho \in (0, \frac{\delta}{3})$ such that if $y \in Q$ and $d(y, Q') < \rho$, then $y \in Q'_\varepsilon$ and, also, if $x \in Q$ and $d(x, Q'_\varepsilon) < \rho$, then $x \in Q'_{2\varepsilon}$. Let n be so large that $D(B, A_n) < \rho$. Take any $x \in B \cap Q'$. We can find $y \in A_n$ such that $d(x, y) < \rho$. However, then $d(y, Q') < \rho$ as well, so $y \in Q'_\varepsilon$. Thus, $B \cap Q'$ is contained in the ρ-neighborhood of $A_n \cap Q'_\varepsilon$.

Assume now that $y \in A_n \cap Q'_\varepsilon$. Then there exists a point $x \in B$ such that $d(x, y) < \rho$. Since $d(x, Q'_\varepsilon) < \rho$, we have $x \in Q'_{2\varepsilon}$. But $B \cap Q'_{2\varepsilon}$ is contained in the $\frac{\delta}{3}$-neighborhood of $B \cap Q'$, so y is contained in the $(\frac{\delta}{3} + \rho)$-neighborhood of $B \cap Q'$. Thus, $D(B \cap Q', A_n \cap Q'_\varepsilon) < \frac{\delta}{3} + \rho < \delta$.

Let now T_ε be the affine mapping of the kind $x \mapsto \lambda x + u$ such that $T_\varepsilon Q'_\varepsilon = Q$. Then T_ε is close to T uniformly on Q when ϵ is close to 0, so $T_\varepsilon(A_n \cap Q'_\varepsilon)$ is still close to $T(B \cap Q')$. However, the latter is a general mini-set of B while the former is always a mini-set of A. $\square$

Our goal is to prove the following

THEOREM 2. *For every fractal A, there exists a micro-set B of A such that $\dim B = \dim^* A$.*

Note that for every mini-set A' of A and every $p > 1$, we have $H_p^*(A') \le H_p^*(A)$. Suppose now that B is a micro-set of A and p is large. Take any cube $Q' \subset Q$ and partition it into p^n equal subcubes. We may choose Q' so that the number of subcubes in this partition that intersect B equals to $H_p^*(B)$.

Choose a sequence A_m of mini-sets of A converging to B. Pick a very small $\varepsilon > 0$. As we have seen above, $B \cap Q'$ is contained in a very small neighborhood of

$A_m \cap Q'_\varepsilon$ if m is large enough. Thus, if B intersects one of the partition subcubes of Q', then A_m must intersect either the corresponding subcube in the partition of Q'_ε or one of its adjacent subcubes. This results in the inequality

$$H_p^*(B) \le 3^n \, H_p^*(A_m) \le 3^n \, H_p^*(A)$$

and

$$\frac{\log H_p^*(B)}{\log p} \le \frac{\log H_p^*(A)}{\log p} + n \, \frac{\log 3}{\log p}.$$

Passing to the limit as $p \to \infty$, we get $dim^* B \le dim^* A$.

Thus, the set B whose existence is claimed in the theorem must satisfy

$$dim^* A = dim \, B \le \underline{M}\text{-}dim \, B \le \overline{M}\text{-}dim \, B \le dim^* B \le dim^* A,$$

i.e.,

$$dim \, B = \underline{M}\text{-}dim \, B = dim^* B = dim^* A.$$

Theorem 2 will follow from Theorem 3 relating to "galleries of trees" formulated in Chapter 9. Theorem 3 in turn implies a generalization of Theorem 2 which deals with the gallery of microsets of a given fractal to arbitrary galleries. In this version we use the star dimension of a gallery Γ defined by

$$dim^* \Gamma = \lim_{p \to \infty} \frac{\log H_p^*(\Gamma)}{\log p} \quad \text{with } H_p^*(\Gamma) = \max_{A \in \Gamma} H_p^*(A).$$

Our main theorem now takes the form:

Theorem 2′. In every gallery Γ, there exists a fractal B with $dim B = dim^* \Gamma$.

We will prove the theorem in several steps. Our first step will be to associate with each fractal $A \subset Q \subset \mathbb{R}^n$ a tree in the same way as we did for one-dimensional sets. The only difference is that now we will use the alphabet $\Lambda = \{0, 1, \ldots, p-1\}^n$ and, given a word $w = \xi_1 \ldots \xi_k \in \Lambda^*$, will associate with it the cube Q_w of sidelength $\frac{1}{p^k}$ whose "left bottom" corner is $\sum_{j=1}^{k} \frac{\xi_j}{p^j}$. We leave it to the reader as an exercise to show that the identity

$$dim \, \widehat{\tau}_A = (\log p) \cdot dim \, A,$$

proved in Lemma 3 of Chapter 3 for one-dimensional sets, still holds in this n-dimensional setting.

CHAPTER 7

Probability Trees Revisited

We shall now present a slightly different angle from which to view probability trees. Consider the set $\Lambda^{\mathbb{N}}$ of all infinite words in the alphabet Λ. It can be viewed as a compact topological space with the usual infinite product topology (where each copy of Λ is endowed with the discrete topology).

Let now $\mathfrak{P}$ be the semi-ring of all cylinders $P_w = \{\omega \in \Lambda^{\mathbb{N}} : \omega = w\omega'\}$ with $w \in \Lambda^*$.

Put $m_\Theta(P_w) = \Theta(w)$.

CLAIM 4. m_Θ is countably additive on $\mathfrak{P}$.

PROOF. Assume that P_w can be written as a disjoint union $\bigcup_j P_{w_j}$, $w, w_j \in \Lambda^*$. Since P_w is closed and P_{w_j} are open in the topology of $\Lambda^{\mathbb{N}}$, this partition must actually be finite. Since $P_{w_j} \subset P_w$, the word w must coincide with an initial segment of w_j, so $w_j = wu_j$ for some $u_j \in \Lambda^*$. Since no infinite word $\omega \in P_w$ can escape all P_{w_j}, we conclude that every word u of length $l(u) \geq \max_j l(u_j)$ must have one of the u_j as its initial segment, so the set $S = \{u_j\}_j$ is a section of Λ^*. Moreover, this section is minimal because if $u_{j'}$ is an initial segment of some $u_{j''}$ with $j'' \neq j'$, then $P_{w_{j''}} \subset P_{w_{j'}}$, which contradicts the assumption that the union is disjoint. Now we consider two cases:

Case 1: $\Theta(w) = 0$. Then $\Theta(w') = 0$ for every word w' starting with w, so $\Theta(w_j) = 0$ for all j and the equality $m_\Theta(P_w) = \sum_j m_\Theta(P_{w_j})$ holds trivially.

Case 2: $\Theta(w) > 0$. Then, $\widetilde{\Theta}(u) = \frac{\Theta(wu)}{\Theta(w)}$ is a probability tree, so, since S is a minimal section of Λ^*, we conclude that

$$1 = \sum_{u \in S} \widetilde{\Theta}(u) = \sum_j \widetilde{\Theta}(u_j)$$

whence (after the multiplication by $\Theta(w)$), $m_\Theta(P_w) = \sum_j m_\Theta(P_{w_j})$. $\qquad\square$

By the Caratheodory theorem, (see [**KF**], Chapter 7, 27), m_Θ extends from $\mathfrak{P}$ to the Borel σ-algebra on $\Lambda^{\mathbb{N}}$. We will denote this measure by the same symbol m_Θ.

Let now $\tau = \tau_\Theta = \{w \in \Lambda^* : \Theta(w) > 0\}$ be the tree associated with the probability tree Θ. Note that every word $\omega \in \Lambda^{\mathbb{N}} \setminus \partial\tau$ has an initial segment $w \in \Lambda^*$ with $\Theta(w) = 0$. Thus, $\Lambda^{\mathbb{N}} \setminus \partial\tau = \bigcup_{w \in \Lambda^*, \Theta(w)=0} P_w$. Since $m_\Theta(P_w) = 0$ whenever $\Theta(w) = 0$ and since Λ^* is countable, we conclude that $m_\Theta(\Lambda^{\mathbb{N}} \setminus \partial\tau) = 0$.

We will now relax our main sufficient condition for the estimation of the dimension of a tree from below.

PROPOSITION 3. *Assume that Θ is a probability tree and*

$$\tau = \tau_\Theta = \{w \in \Lambda^* : \Theta(w) > 0\}$$

is the associated tree. Let m_Θ be the probability measure constructed above. Assume that the set of words $\omega \in \Lambda^{\mathbb{N}}$ for which

$$\limsup_{w \to \omega} \frac{\log \Theta(w)}{l(w)} \leq -\lambda$$

has positive m_Θ measure (here the upper limit is taken over the sequence of all initial segments of ω and $\log 0 = -\infty$). Then $\dim \tau_\Theta \geq \lambda$.

Before proving this proposition, let us note that, since $\partial \tau_\Theta$ has full measure, we can just as well write "for m_Θ-almost every $\omega \in \partial \tau_\Theta$" instead.

PROOF. Let $\lambda' < \lambda$. Observe that

$$\{\omega \in \Lambda^{\mathbb{N}} : \limsup_{w \to \omega} \frac{\log \Theta(w)}{l(w)} \leq -\lambda\} \subset \bigcup_{L>1} E_L$$

where

$$E_L = \{\omega \in \Lambda^{\mathbb{N}} : \Theta(w) \leq e^{-\lambda' l(w)} \quad \text{for all initial segments } w \text{ of } \omega \text{ with } l(w) \geq L\}.$$

Thus, there exists L such that $m_\Theta(E_L) > 0$. Now take any section S of τ_Θ. If S contains a word of length L or less, then $\sum_{w \in S} e^{-\lambda' l(w)} \geq e^{-\lambda' L}$. Otherwise note that $\partial \tau_\Theta \subset \bigcup_{w \in S} P_w$, so

$$m_\Theta(E_L) = m_\Theta(E_L \cap \partial \tau_\Theta) \leq \sum_{w \in S, \, P_w \cap E_L \neq \varnothing} m_\Theta(P_w).$$

However, $P_w \cap E_L \neq \varnothing$ only if there exists a word $\omega \in E_L$ whose initial segment is w. Since $l(w) \geq L$, we must have $m_\Theta(P_w) = \Theta(w) \leq e^{-\lambda' l(w)}$ and we finally conclude that in this case $\sum_{w \in S} e^{-\lambda' l(w)} \geq m_\Theta(E_L)$.

In both cases, we have a positive lower bound for $\sum_{w \in S} e^{-\lambda' l(w)}$ independent of the section S. Thus, $\dim \tau_\Theta \geq \lambda'$. Since $\lambda' < \lambda$ was arbitrary, we conclude that $\dim \tau_\Theta \geq \lambda$. $\qquad\square$

With every probability tree Θ, one can associate the *Markov process* whose set of states is τ_Θ and whose transition probabilities are

$$p_{uv} = \begin{cases} \frac{\Theta(v)}{\Theta(u)} & \text{if } u, v \in \tau_\Theta \text{ and } v = ux, \ x \in \Lambda; \\ 0 & \text{otherwise.} \end{cases}$$

In layman's terms, this Markov process is just a randomly growing word that starts empty and if one has word u at some point, then the probability to add a letter x to this word at the next step equals $\frac{\Theta(ux)}{\Theta(u)}$. From this point of view, the measure $m_\Theta(E)$ of a Borel set $E \subset \Lambda^{\mathbb{N}}$ is just the probability that this growing word will end up in E.

Elements of Ergodic Theory

Let Ω be a set, let $T : \Omega \to \Omega$ be any transformation from Ω to itself, and let f be a real-valued function on Ω. Define *ergodic averages* $A_N f$ by

$$(A_N f)(x) = \frac{1}{N} \sum_{j=0}^{N-1} f(T^j x).$$

Assume in addition that we have a probability measure μ on some σ-algebra $\mathcal{B}$ of subsets of Ω and that T is measurable with respect to $\mathcal{B}$ (i.e., $T^{-1} B \in \mathcal{B}$ for all $B \in \mathcal{B}$) and *measure preserving* (i.e., $\mu(T^{-1} B) = \mu(B)$ for all $B \in \mathcal{B}$).

The *Birkhoff ergodic theorem* asserts that if f is integrable with respect to μ, then $\lim_{N \to \infty} (A_N f)(x) = \bar{f}(x)$ exists for μ-almost all x and $\int_\Omega \bar{f} d\mu = \int_\Omega f d\mu$. The reader can find the proof of the Birkhoff ergodic theorem in ([**K**], page 7).

Recall that a *stationary random process* is just a sequence of random variables X_j on some probability space with values in some set M such that for every $k \geq 0$ and every (measurable) $B \subset M^k$, we have

$$\mathbb{P}\{(X_0, \ldots, X_{k-1}) \in B\} = \mathbb{P}\{(X_m, \ldots, X_{m+k-1}) \in B\}$$

for all $m \geq 0$. If we consider $(X, \mathcal{B}, \mu)$ as a probability space, then for every measurable measure-preserving transformation $T : \Omega \to \Omega$ and every measurable function $f : \Omega \to M$, the sequence $\Omega_j = f \circ T^j$ is a stationary process. The proof is straightforward:

$$\mathbb{P}\{(X_m, \ldots, X_{m+k-1}) \in B\} = \mu\{x \in X : (f(T^m x), \ldots, f(T^{m+k-1} x)) \in B\}$$

$$= \mu(T^{-m}\{x \in X : (f(x), \ldots, f(T^{k-1} x)) \in B\})$$

$$= \mu(\{x \in X : (f(x), \ldots, f(T^{k-1} x)) \in B\}) = \mathbb{P}\{(X_0, \ldots, X_{k-1}) \in B\}.$$

In what follows, we shall take M to be the space of all probability trees over some alphabet Λ with the natural topology induced from the product topology on $\mathbb{R}^{\Lambda^*}$ viewed as the space of real-valued functions on Λ^*.

For a probability tree Θ and a word $u \in \Lambda^*$ with $\Theta(u) > 0$, we will denote by Θ^u the probability tree defined by

$$\Theta^u(w) = \frac{\Theta(uw)}{\Theta(u)} \qquad (w \in \Lambda^*).$$

Recall that to estimate the dimension of τ_Θ from below, we may look at the limit $\lim_{w \to \omega} \frac{\log \Theta(w)}{l(w)}$. Our next aim is to represent the quantity under the limit sign as an ergodic average. To this end, we will consider the space X consisting of all pairs $x = (\Theta, \omega)$ where Θ is a probability tree on Λ^* and $\omega = \xi_1 \xi_2 \cdots \in \partial \tau_\Theta$. On this space, we will define the transformation T by $T(\Theta, \xi_1 \xi_2 \ldots) = (\Theta^{\xi_1}, \xi_2 \xi_3 \ldots)$ and the function $f((\Theta, \xi_1 \xi_2 \ldots)) = \log \Theta(\xi_1)$.

Now $T^j(\Theta, \xi_1\xi_2\dots) = (\Theta^{\xi_1\dots\xi_j}, \xi_{j+1}\xi_{j+2}\dots)$, so

$$f(T^j(\Theta, \xi_1\xi_2\dots)) = \log \Theta^{\xi_1\dots\xi_j}(\xi_{j+1}) = \log \frac{\Theta(\xi_1\dots\xi_{j+1})}{\Theta(\xi_1\dots\xi_j)},$$

whence

$$\frac{1}{N}\sum_{j=0}^{N-1} f(T^j(\Theta, \xi_1\xi_2\dots)) = \frac{1}{N}\sum_{j=0}^{N-1} \log \frac{\Theta(\xi_1\dots\xi_{j+1})}{\Theta(\xi_1\dots\xi_j)} = \frac{1}{N}\log \Theta(\xi_1\dots\xi_N).$$

The general idea is that if we could apply the ergodic theorem now and show that this quantity converges to a limit, we would have some information about the Hausdorff dimension of something. The task is just to ensure that this information and that something are exactly what we need.

CHAPTER 9

Galleries of Trees

Recall that our aim is to prove (using Ergodic Theory) that for every fractal A, one can find a micro-set B of A such that $dim\, B = dim^*\, A$. Since the set of all micro-sets of a given fractal A is a gallery, it suffices to prove that for every gallery Γ of fractals, there exists $B \in \Gamma$ such that $dim\, B = dim^*\, \Gamma$ where $dim^*\, \Gamma$ is defined as

$$dim^*\, \Gamma = \lim_{p \to \infty} \frac{\log H_p^*(\Gamma)}{\log p} \qquad \text{with} \qquad H_p^*(\Gamma) = \max_{A \in \Gamma} H_p^*(A).$$

The proof that the limit exists is almost identical to that of Lemma 2 in Chapter 2. Again, we would like to pass from sets to trees. That will require us to define what a gallery of trees and its star-dimension are.

DEFINITION. *Let τ be a tree and $w \in \tau$ be a word. Define the successor tree τ^w corresponding to the word w by*

$$\tau^w = \{u \in \Lambda^* : wu \in \tau\}.$$

We will also introduce the topology on the set of all trees. The basic neighborhood $U_L(\tau)$, $L \geq 0$, of a tree τ in this topology will be defined as the set of all trees τ' such that a word $w \in \Lambda^*$ of length $l(w) \leq L$ belongs to τ' if and only if it belongs to τ (i.e., τ' coincides with τ up to length L, after which it is unrestricted).

DEFINITION. *A set $\widehat{\Gamma}$ of trees is a gallery of trees if*
1) Every successor tree of every tree in $\widehat{\Gamma}$ is again in $\widehat{\Gamma}$.
2) $\widehat{\Gamma}$ is closed in the topology introduced above.

Our next goal is to show that with every gallery of fractals, one can associate a gallery of trees in some natural way so that the theorem we want to prove will be equivalent to the corresponding theorem for trees.

Since we have already associated a tree $\widehat{\tau}_A$ with every fractal A, the naive idea would be just to define the gallery of trees associated with a gallery Γ of fractals by $\widehat{\Gamma} = \widehat{\Gamma}_0 = \{\widehat{\tau}_A : A \in \Gamma\}$.

Unfortunately, it is not immediately clear why $\widehat{\Gamma}_0$ must be closed (the difficulty is the same as in Chapter 6: in general $\lim_{n \to \infty} (A_n \cap Q) \neq (\lim_{n \to \infty} A_n) \cap Q$), so instead of trying to figure out if $\widehat{\Gamma}_0$ is always closed or not, we will just take the closure on the right hand side and define $\widehat{\Gamma} = Cl\, \widehat{\Gamma}_0$.

Note that if $\tau \in \widehat{\Gamma}_0$, then every successor τ^w of τ is again in $\widehat{\Gamma}_0$. To see it, just note that $\tau^w = \widehat{\tau}_{A'}$, where A' is the mini-set of A corresponding to the cube Q_w associated with the word w.

35

Suppose now that $\tau \in \widehat{\Gamma}$ and $w \in \tau$.

Take any basic neighborhood $U_L(\tau^w)$. Take any $\tilde{\tau} \in \widehat{\Gamma}_0 \cap U_{L+l(w)}(\tau)$. Then, clearly, $\tilde{\tau}^w \in \widehat{\Gamma}_0 \cap U_L(\tau^w)$. Thus, $\tau^w \in Cl\,\widehat{\Gamma}_0$ whenever $\tau \in Cl\,\widehat{\Gamma}_0$, so every successor of every tree in $\widehat{\Gamma}$ is again in $\widehat{\Gamma}$ and we conclude that $\widehat{\Gamma}$ is a gallery of trees.

DEFINITION. *The star dimension $dim^* \Gamma$ of a gallery Γ of trees is the limit* $\lim\limits_{l \to \infty} \frac{\log N_l^*(\Gamma)}{l}$ *where* $N_l^*(\Gamma) = \max\limits_{\tau \in \Gamma} N_l(\tau)$.

Again, the proof of the existence of this limit is very similar to that in Lemma 2 of Chapter 2, so we leave it to the reader as an (easy) exercise.

CLAIM 5. $dim^* \widehat{\Gamma} = (dim^* A) \log p$.

PROOF. It follows immediately from the definition that $N_l(\widehat{\tau}_A) \leq H_{p^l}^*(A)$ for every fractal A. This allows us to conclude that $N_l(\tau) \leq H_{p^l}^*(\Gamma)$ for all $\tau \in \widehat{\Gamma}_0$ and, thereby, after passing to the limit, for all $\tau \in \widehat{\Gamma}$. Thus, $N_l^*(\widehat{\Gamma}) \leq H_{p^l}^*(\Gamma)$ and $dim^* \widehat{\Gamma} \leq (dim^* A) \log p$.

To prove the reverse inequality, take any set $A \subset \Gamma$ and any cube $Q' \subset Q$ divided into $(p^l)^n$ equal subcubes. Note that (by passing to a mini-set, if necessary), we can always assume that $Q' = Q$. Then $N_{p^l}(A) = N_l(\widehat{\tau}_A)$, so

$$H_{p^l}^*(\Gamma) = \max_{A \in \Gamma} H_{p^l}^*(A) = \max_{A \in \Gamma} N_{p^l}(A) \leq \max_{\tau \in \widehat{\Gamma}_0} N_l(\tau) \leq \max_{\tau \in \widehat{\Gamma}} N_l(\tau) = N_l^*(\widehat{\Gamma}).$$

Taking logarithms, dividing by l, and passing to the limit, we get the desired result.   □

The tree version of our main theorem reads as follows:

THEOREM 3. *Let $\widehat{\Gamma}$ be a gallery of trees. Then there exists $\tau \in \widehat{\Gamma}$ such that $dim\,\tau = dim^* \widehat{\Gamma}$.*

Applying this theorem to the gallery of trees $\widehat{\Gamma}$ associated with a gallery Γ of sets, we get a tree $\tau \in \widehat{\Gamma}$ whose Hausdorff dimension is $dim\,\tau = (dim^* \Gamma) \log p$.

All we need to show is that this tree corresponds to some set in Γ.

CLAIM 6. *If Γ is a gallery of sets and $\widehat{\Gamma}$ is the associated gallery of trees, then for every $\tau \in \widehat{\Gamma}$, there exists $A \in \Gamma$ such that $\tau \subset \widehat{\tau}_A$.*

PROOF. Observe that for every $L \geq 0$, there exists $A_L \in \Gamma$ such that $\widehat{\tau}_{A_L} \in U_L(\tau)$.

If A_L had converged to some limit A in the Hausdorff metric, then we would have $\widehat{\tau}_A \supset \tau$. In general, the limit of A_L does not exist but, since the space of all closed subsets of Q endowed with the Hausdorff distance is compact (see [**Fe**], page 183), we can pass to a converging subsequence and get the same result.   □

So

$$dim\,A = \frac{dim\,\widehat{\tau}_A}{\log p} \geq \frac{dim\,\tau}{\log p} = dim^* \Gamma.$$

Since we always have $dim\,A \leq dim^* \Gamma$ for every $A \in \Gamma$, we would have exactly what we wanted in this case.

From now on, we will almost never mention fractals again, but concentrate on trees instead.

CHAPTER 10

General Remarks on Markov Systems

Let $\mathfrak{M}$ be a compact topological space. A Markov process on $\mathfrak{M}$ is defined by its transition probability function $p : \mathfrak{M} \times \mathfrak{M} \to [0, 1]$ where for every $x \in \mathfrak{M}$, the value $p(x, y) > 0$ only for finitely many $y \in \mathfrak{M}$ and $\sum_{y \in \mathfrak{M}} p(x, y) = 1$.

The quantity $p(x, y)$ can be interpreted as the probability to arrive at the state $y \in \mathfrak{M}$ at the next moment if we are at the state x at the current moment.

Define the *Markov operator* $\mathcal{T}$ on the space of continuous functions by

$$\mathcal{T} f(x) = \sum_{y \in \mathfrak{M}} p(x, y) f(y).$$

Of course, the condition that $\mathcal{T} f$ is continuous whenever f is is quite a severe restriction on p but it will be satisfied in all our examples.

For a Borel probability measure μ on $\mathfrak{M}$, define $\mathcal{T}^* \mu$ as the unique Borel measure on $\mathfrak{M}$ satisfying

$$\int_{\mathfrak{M}} f d(\mathcal{T}^* \mu) = \int_{\mathfrak{M}} (\mathcal{T} f) d\mu$$

for all continuous $f : \mathfrak{M} \to \mathbb{R}$ (it exists according to the Riesz theorem).

We call the measure μ stationary if $\mathcal{T}^* \mu = \mu$.

Suppose that μ is stationary. Define the Markov process $X_0, X_1, \ldots$ with values in $\mathfrak{M}$ as follows: X_0 is distributed according to μ. At each moment, the conditional probability that $X_{n+1} = y$ under the condition $X_n = x$ equals $p(x, y)$ (i.e., at each step, we jump from the current state x to a new state y with probability $p(x, y)$).

Note that this definition may face severe problems with measurability in general. However, we will use it only as a guideline for the actual rigorous argument, so we will sweep all such issues under the rug for now.

The point is that the process $X_1, X_2, \ldots$ defined in this way is stationary. To see it, note first that if μ_j is the distribution of X_j (i.e., $\mathbb{P}(X_j \in B) = \mu_j(B)$), then $\mu_{j+1} = \mathcal{T}^* \mu_j$, so in the case when μ is stationary, all X_j have the same distribution. However, due to the Markov property, the joint distribution of X_m, $X_{m+1}, \ldots, X_{m+k-1}$ is completely determined by the distribution of X_m and the transition probabilities $p(x, y)$, which do not change from one step to another.

CHAPTER 11

Markov Operator $\mathcal{T}$ and Measure Preserving Transformation T

Let us now pass to the formal argument.

Let $\mathcal{P}$ be the set of all probability trees. We want to introduce the natural Markov process on $\mathcal{P}$, which moves $\Theta \in \mathcal{P}$ to Θ^a ($a \in \Lambda$) with probability $\Theta(a)$ at each step. The corresponding Markov operator $\mathcal{T}$ in this case will be given by

$$(\mathcal{T}f)(\Theta) = \sum_{a \in \Lambda} \Theta(a) f(\Theta^a)$$

and its m-th power will be given by

$$(\mathcal{T}^m f)(\Theta) = \sum_{w \in \Lambda^*, \, l(w)=m} \Theta(w) f(\Theta^w).$$

Note that Θ^a (or Θ^w) is not defined when $\Theta(a)$ (or $\Theta(w)$) vanishes. However, since the undefined value of f is multiplied by 0 at this case, we can just ignore the corresponding terms.

Now we will change our usual notation and denote by $\xi = \xi_1 \xi_2 \xi_3 \ldots$ infinite words in the alphabet Λ reserving the letter ω for the elements of the probability space we are about to construct.

Let us call a measure μ on $\mathcal{P}$ stationary if for every continuous function $f : \mathcal{P} \to \mathbb{R}$, we have

$$\int_{\mathcal{P}} f(\Theta) d\mu(\Theta) = \int_{\mathcal{P}} \sum_{a \in \Lambda} \Theta(a) f(\Theta^a) d\mu(\Theta)$$

(note that for every $a \in \Lambda$, the mapping $\Theta \mapsto \Theta(a) f(\Theta^a)$ extended by 0 to the set where $\Theta(a) = 0$ is continuous, so the integral on the right is well-defined).

We now want to define a Borel probability measure $\mathbb{P} = \mathbb{P}_\mu$ on

$$\Omega = \mathcal{P} \times \Lambda^{\mathbb{N}} = \{\omega = (\Theta, \xi) : \Theta \in \mathcal{P}, \xi \in \Lambda^{\mathbb{N}}\}$$

by

$$\int_{\Omega} F(\Theta, \xi) d\mathbb{P} = \int_{\mathcal{P}} \left(\int_{\Lambda^{\mathbb{N}}} F(\Theta, \xi) dm_\Theta(\xi) \right) d\mu(\Theta).$$

If the functional on the right is well-defined, it is, clearly, linear and non-negative on the class of continuous functions $F : \Omega \to \mathbb{R}$. So to apply the Riesz theorem[1] and establish the existence of a unique measure $\mathbb{P}$ with this property, it suffices to show that the repeated integral on the right makes sense.

To this end, we will demonstrate that the mapping $\Theta \mapsto \int_{\Lambda^{\mathbb{N}}} F(\Theta, \xi) dm_\Theta(\xi)$ is

continuous.

---

[1] Note that $\mathcal{P}$ and Ω are compact topological spaces in their natural topologies.

Note that, by the Stone-Weierstrass Theorem, every continuous function $F(\Theta, \xi)$ of two variables can be uniformly approximated by finite linear combinations of elementary functions $F(\Theta, \xi) = G(\Theta)H(\xi)$ where G is continuous on $\mathcal{P}$ and H is continuous on $\Lambda^{\mathbb{N}}$. Moreover, performing one more approximation procedure, if necessary, we may assume that H depends only on the first L symbols of ξ for some (large) L. But for such functions, we have

$$\int_{\Lambda^{\mathbb{N}}} F(\Theta, \xi) dm_{\Theta}(\xi) = G(\Theta) \sum_{w \in \Lambda^*, \, l(w) = L} H(w) \Theta(w),$$

which is, clearly, continuous in Θ. It remains to recall that the integration with respect to a probability measure does not increase the size of the approximation error and that the uniform limit of continuous functions is continuous.

Consider the transformation $T\omega = (\Theta^{\xi_1}, \sigma\xi)$ on Ω, where σ is the backward shift on $\Lambda^{\mathbb{N}}$ defined by $\sigma(\xi_1 \xi_2 \xi_3 \dots) = \xi_2 \xi_3 \xi_4 \dots$.

CLAIM 7. *The transformation T is defined $\mathbb{P}$-almost everywhere.*

PROOF. The problematic ω are those for which $\Theta(\xi_1) = 0$. However, we will now show that even the set of ω for which $\xi \notin \partial \tau_{\Theta}$ is a null-set.

To this end, let $m, k > 0$ and consider the continuous functions

$$F_{mk}(\Theta, \xi) = (1 - \Theta(\xi^{(k)}))^m$$

where $\xi^{(k)}$ is the initial segment of ξ of length k. We always have $0 \leq F_{mk} \leq 1$. Also, for a fixed k,

$$F_{mk}(\Theta, \xi) \to F_k(\Theta, \xi) = \begin{cases} 1, & \Theta(\xi^{(k)}) = 0, \\ 0 & \text{otherwise} \end{cases}$$

as $m \to \infty$. On the other hand,

$$\int_{\Omega} F_{mk} d\mathbb{P} = \int_{\mathcal{P}} \left(\sum_{w \in \Lambda^*, \, l(w) = k} \Theta(w)(1 - \Theta(w))^m \right) d\mu(\Theta).$$

However, the functions $t \mapsto t(1-t)^m$ tend to 0 uniformly on $[0, 1]$ and we have only $|\Lambda|^k$ words of length k, so $\int_{\Omega} F_{mk} d\mathbb{P} \to 0$ as $m \to \infty$.

The conclusion is that $\mathbb{P}\{\omega : \Theta(\xi^{(k)}) = 0\} = 0$. However,

$$\{\omega : \xi \notin \partial \tau_{\Theta}\} = \bigcup_{k=0}^{\infty} \{\omega : \Theta(\xi^{(k)}) = 0\},$$

so it is also a null-set with respect to the measure $\mathbb{P}$. $\square$

CLAIM 8. *Assume that the measure μ is stationary. Then T is measure-preserving.*

PROOF. It suffices to show that

$$\int_{\Omega} \left(F \circ T \right) d\mathbb{P} = \int_{\Omega} F d\mathbb{P}$$

for every continuous function $F : \Omega \to \mathbb{R}$. Note that in general $F \circ T$ is worse than F: it is not even defined everywhere. So, when applying the definition of $\mathbb{P}$ by duality, one needs to be somewhat careful.

By the definition of stationarity, we have

$$\int\limits_{\Omega} F d\mathbb{P} = \int\limits_{\mathcal{P}} \Big[\sum_{a \in \Lambda} \Theta(a) \int\limits_{\Lambda^{\mathbb{N}}} F(\Theta^a, \xi) dm_{\Theta^a}(\xi) \Big] d\mu.$$

To evaluate $\int\limits_{\Omega} \Big(F \circ T \Big) d\mathbb{P}$, we start with noticing that, although the formula

$$\int\limits_{\Omega} F d\mathbb{P} = \int\limits_{\mathcal{P}} \Big[\int\limits_{\Lambda^{\mathbb{N}}} F(\Theta, \xi) dm_{\Theta}(\xi) \Big] d\mu(\Theta)$$

is initially assumed to hold only for continuous F, it can be easily extended to the class of all bounded Borel measurable F.

To do it, observe that if we have a pointwise converging sequence of uniformly bounded functions F_n for which the formula holds (i.e., the right hand side makes sense and the equality holds), then, applying the dominated convergence theorem three times (once on the left and twice on the right), we get equality for the limiting function $F = \lim\limits_{n \to \infty} F_n$.

However, the least class of functions on Ω that contains all continuous functions and is closed under taking pointwise limits of uniformly bounded sequences of functions is the class of all bounded Borel measurable functions. Using the monotone convergence theorem instead of the dominated convergence theorem, we can extend this class to that of all *integrable* Borel measurable functions (possibly with infinite integrals).

Now, noting that T is well-defined and continuous on the open set $\Omega' = \{\omega : \Theta(\xi_1) > 0\}$, we conclude that $F \circ T$ (extended, say, by zero to the set $\Omega \setminus \Omega'$) is bounded and Borel measurable, for every continuous $F : \Omega \to \mathbb{R}$.

Hence,

$$\int\limits_{\Omega} \Big(F \circ T \Big) d\mathbb{P} = \int\limits_{\mathcal{P}} \Big[\int\limits_{\Lambda^{\mathbb{N}}} F(\Theta^{\xi_1}, \sigma\xi) dm_{\Theta}(\xi) \Big] d\mu(\Theta)$$

$$= \int\limits_{\mathcal{P}} \Big[\sum_{a \in \Lambda} \int\limits_{\xi \in \Lambda^{\mathbb{N}} : \xi_1 = a} F(\Theta^a, \sigma\xi) dm_{\Theta}(\xi) \Big] d\mu(\Theta).$$

Note again that for Θ such that $\Theta(a) = 0$, the value $F(\Theta^a, \sigma\xi)$ is undefined but

$$m_{\Theta}\{\xi \in \Lambda^{\mathbb{N}} : \xi_1 = a\} = \Theta(a) = 0,$$

so the inner integral still can be evaluated as 0. If $\Theta(a) > 0$, the inner integral makes perfect sense and we can write

$$\int\limits_{\xi \in \Lambda^{\mathbb{N}} : \xi_1 = a} F(\Theta^a, \sigma\xi) dm_{\Theta}(\xi) = \Theta(a) \int\limits_{\Lambda^{\mathbb{N}}} F(\Theta^a, \xi') dm_{\Theta^a}(\xi')$$

because the change of variable $\varphi : \xi \mapsto \xi' = \sigma\xi$ is a continuous bijection with continuous inverse from $\{\xi \in \Lambda^{\mathbb{N}} : \xi_1 = a\}$ onto $\Lambda^{\mathbb{N}}$ and we have $m_{\Theta}(E) = \Theta(a) m_{\Theta^a}(\varphi(E))$ for every Borel set E (this is obvious from the definition of Θ^a for every element P_w of the semiring $\mathfrak{P}$, but the extension of a measure on $\mathfrak{P}$ to a Borel measure is unique). Thus, the inner integral is always equal to $\Theta(a) \int\limits_{\Lambda^{\mathbb{N}}} F(\Theta^a, \xi) dm_{\Theta^a}(\xi)$ (with the usual interpretation of "0 times undefined" as 0) and we are done. $\qquad\square$

Let us now consider the *Entropy function*

$$\mathcal{E}(\Theta) = -\sum_{a \in \Lambda} \Theta(a) \log \Theta(a)$$

and the *Information function*

$$\mathcal{I}(\omega) = -\log \Theta(\xi_1).$$

Note that

$$(2) \qquad \int_{\Omega} \mathcal{I} d\mathbb{P} = \int_{\mathcal{P}} \left(\int_{\Lambda^{\mathbb{N}}} -\log \Theta(\xi_1) dm_{\Theta}(\xi) \right) d\mu(\Theta) = \int_{\mathcal{P}} \mathcal{E}(\Theta) d\mu(\Theta).$$

We will also view $\mathcal{E}$ as a function on Ω (depending on the first variable only).

Let us apply the Markov operator $\mathcal{T}$ to $\mathcal{E}$ and look at the average $\frac{1}{L} \sum_{j=0}^{L-1} \mathcal{T}^j \mathcal{E}$.

Note that for an arbitrary continuous $F : \Omega \to \mathbb{R}$, we have

$$\mathcal{T}^j F = \sum_{w \in \Lambda^*, \, l(w)=j} \Theta(w) F(\Theta^w).$$

Also,

$$\mathcal{E}(\Theta^w) = -\sum_{a \in \Lambda} \Theta^w(a) \log \Theta^w(a) = -\sum_{a \in \Lambda} \frac{\Theta(wa)}{\Theta(w)} \log \frac{\Theta(wa)}{\Theta(w)},$$

so

$$\sum_{w \in \Lambda^*, \, l(w)=j} \Theta(w) \mathcal{E}(\Theta^w) = -\sum_{a \in \Lambda, \, w \in \Lambda^*, \, l(w)=j} \Theta(wa)[\log \Theta(wa) - \log \Theta(w)]$$

$$= -\sum_{w \in \Lambda^*, \, l(w)=j+1} \Theta(w) \log \Theta(w) + \sum_{w \in \Lambda^*, \, l(w)=j} \Theta(w) \log \Theta(w).$$

Thus, taking into account that $\Theta(\varnothing) = 1$, so $\Theta(\varnothing) \log \Theta(\varnothing) = 0$, we conclude that

$$\frac{1}{L} \sum_{j=0}^{L-1} \mathcal{T}^j \mathcal{E}(\Theta) = -\frac{1}{L} \sum_{w \in \Lambda^*, \, l(w)=L} \Theta(w) \log \Theta(w).$$

CHAPTER 12

Probability Trees and Galleries

Let now $\widehat{\Gamma}$ be a gallery of trees.

Consider the subset $\mathcal{P}_{\widehat{\Gamma}} = \{\Theta \in \mathcal{P} : \tau_\Theta \subset \tau \text{ for some } \tau \in \widehat{\Gamma}\}$. Note that the definition of a gallery implies that $\mathcal{P}_{\widehat{\Gamma}}$ is a closed subset of $\mathcal{P}$ and thereby, compact, and that $\Theta^a \in \mathcal{P}_{\widehat{\Gamma}}$ whenever $\Theta \in \mathcal{P}_{\widehat{\Gamma}}$ and Θ^a is defined.

To see the first statement, assume that some sequence of probability trees $\Theta_n \in \mathcal{P}$ converges to $\Theta \in \mathcal{P}$ (note that $\mathcal{P}$ is metrizable, so to show that a subset of $\mathcal{P}$ is closed, it suffices to show that it contains all limits of sequences consisting of its elements). Let $\tau_{\Theta_n} \subset \tau_n \in \widehat{\Gamma}$. By compactness of the space of trees, we assume (passing to a subsequence if necessary) that τ_n converge to some tree $\tau \in \widehat{\Gamma}$.

Now note that if $\Theta(w) > 0$, then $\Theta_n(w) > 0$ for all sufficiently large n, so $w \in \tau_{\Theta_n} \subset \tau_n$. It follows that $\tau_\Theta \subset \tau$, so $\Theta \in \mathcal{P}_{\widehat{\Gamma}}$.

The second statement is immediately implied by the observation that τ_{Θ^a} is a successor tree of τ_Θ and a successor tree of a larger tree is not smaller (if the reader forgot what a successor tree is he can find the definition at the beginning of Chapter 9).

Thus, we can carry out all the analysis in the previous chapters with $\mathcal{P}$ replaced by $\mathcal{P}_{\widehat{\Gamma}}$.

Let $\mathcal{T}$ be the corresponding Markov operator $\mathcal{T}F(\Theta) = \sum_{a \in \Lambda} \Theta(a) F(\Theta^a)$ on the space of continuous functions on $\mathcal{P}_{\widehat{\Gamma}}$ and let T be the transformation $\omega = (\Theta, \xi) \mapsto (\Theta^{\xi_1}, \sigma\xi)$ on $\mathcal{P}_{\widehat{\Gamma}} \times \Lambda^{\mathbb{N}}$.

We can again consider a Borel probability measure μ on $\mathcal{P}_{\widehat{\Gamma}}$ and define the measure $\mathbb{P}$ on $\mathcal{P}_{\widehat{\Gamma}} \times \Lambda^{\mathbb{N}}$ in the same way as before.

The result that if μ is stationary, then T is measure-preserving with respect to $\mathbb{P}$, can also be transferred to this new setting without any changes in the proof.

Assuming that μ is stationary, we have

$$\int_{\mathcal{P}_{\widehat{\Gamma}}} \mathcal{E} d\mu = \frac{1}{L} \sum_{j=0}^{L-1} \int_{\mathcal{P}_{\widehat{\Gamma}}} \left(\mathcal{T}^j \mathcal{E}\right) d\mu = \int_{\mathcal{P}_{\widehat{\Gamma}}} \left(\frac{1}{L} \sum_{j=0}^{L-1} \mathcal{T}^j \mathcal{E}\right) d\mu$$

$$= \int_{\mathcal{P}_{\widehat{\Gamma}}} \frac{1}{L} \sum_{w \in \Lambda^*, \, l(w)=L} \Theta(w)(-\log \Theta(w)) \, d\mu(\Theta).$$

However, we now know that the number of words w of length L for which $\Theta(w) > 0$ does not exceed $N_L(\tau_\Theta) \leq N_L^*(\widehat{\Gamma})$. Since the function $t \mapsto -t \log t$ is concave, it follows that

$$\sum_{w \in \Lambda^*, \, l(w)=L} \Theta(w)(-\log \Theta(w)) \leq \log N_L^*(\widehat{\Gamma})$$

for every $\Theta \in \mathcal{P}_{\widehat{\Gamma}}$. Thus, we obtain the inequality

$$\int_{\mathcal{P}_{\widehat{\Gamma}}} \mathcal{E}d\mu \leq \frac{\log N_L^*(\widehat{\Gamma})}{L}.$$

Passing to the limit as $L \to \infty$, we conclude that $\int_{\mathcal{P}_{\widehat{\Gamma}}} \mathcal{E}d\mu \leq dim^*(\widehat{\Gamma})$. We also have $\int_{\mathcal{P}_{\widehat{\Gamma}} \times \Lambda^{\mathbb{N}}} \mathcal{I}\, d\mathbb{P} = \int_{\mathcal{P}_{\widehat{\Gamma}}} \mathcal{E}d\mu$, so the same inequality holds for the integral $\int_{\mathcal{P}_{\widehat{\Gamma}} \times \Lambda^{\mathbb{N}}} \mathcal{I}\, d\mathbb{P}$.

LEMMA 5. *There exists a stationary measure μ such that $\int_{\mathcal{P}_{\widehat{\Gamma}}} \mathcal{E}d\mu = dim^* \widehat{\Gamma}$.*

PROOF. Note that for every measure μ on $\mathcal{P}_{\widehat{\Gamma}}$, we can define $\mathcal{T}^*\mu$ as the unique measure such that

$$\int_{\mathcal{P}_{\widehat{\Gamma}}} \mathcal{T}f d\mu = \int_{\mathcal{P}_{\widehat{\Gamma}}} f d\left(\mathcal{T}^*\mu\right)$$

for all continuous functions $f : \mathcal{P}_{\widehat{\Gamma}} \to \mathbb{R}$ where, as before, $\mathcal{T}f(\Theta) = \sum_{a \in \Lambda} \Theta(a)f(\Theta^a)$. Since for $f \equiv 1$, we have $\mathcal{T}f \equiv 1$, we conclude that $\mathcal{T}^*\mu$ is a probability measure whenever μ is. Also, the operator $\mathcal{T}^*$ is, clearly, linear.

The condition that μ is stationary can be rewritten as $\mathcal{T}^*\mu = \mu$.

There is a simple recipe for getting an almost stationary probability measure from an arbitrary probability measure ν: just put $\mu_l = \frac{1}{l} \sum_{j=0}^{l-1} (\mathcal{T}^*)^j \nu$ with large $l \in \mathbb{N}$. The measure μ_l is not stationary. However, the difference $\mathcal{T}^*\mu_l - \mu_l = \frac{1}{l}[(\mathcal{T}^*)^l \nu - \nu]$ has total variation at most $\frac{2}{l}$, so if we take the limit of some weakly convergent subsequence of the sequence μ_l, we will get a probability measure μ satisfying $\mathcal{T}^*\mu = \mu$. Moreover, in this construction we may change the measure ν from one step to another.

We will construct our measure μ as a weak* limit of a subsequence of the sequence $\mu_l = \frac{1}{l} \sum_{j=0}^{l-1} (\mathcal{T}^*)^j \nu_l$ with some carefully chosen probability measures ν_l.

Recall that $dim^* \widehat{\Gamma} = \lim_{l \to +\infty} \frac{\log N_l^*(\widehat{\Gamma})}{l}$ and that $N_l^*(\widehat{\Gamma}) = \max_{\tau \in \widehat{\Gamma}} N_l(\tau)$. Thus, for every l, we can find $\tau_l \in \widehat{\Gamma}$ such that $N_l(\tau_l) = N_l^*(\widehat{\Gamma})$.

Let Θ_l be any probability tree with $\tau_{\Theta_l} \subset \tau_l$ such that $\Theta_l(w) = \frac{1}{N_l(\tau_l)}$ on each word $w \in \tau_l$ of length l (this condition determines $\Theta_l(w)$ for words of length $l(w) \leq l$ uniquely but we usually have a lot of freedom in defining Θ_l on longer words).

Put $\nu_l = \delta_{\Theta_l}$ where δ stands for the Dirac unit point mass. In other words, for $E \subset \mathcal{P}_{\widehat{\Gamma}}$,

$$\nu_l(E) = \begin{cases} 1, & \Theta_l \in E \\ 0 & \text{otherwise.} \end{cases}$$

Let us evaluate $\int_{\mathcal{P}_{\widehat{\Gamma}}} \mathcal{E} d\mu_l$. It equals

$$\int_{\mathcal{P}_{\widehat{\Gamma}}} \mathcal{E} d\Big(\frac{1}{l}\sum_{j=0}^{l-1}(\mathcal{T}^*)^j \nu_l\Big) = \int_{\mathcal{P}_{\widehat{\Gamma}}} \Big(\frac{1}{l}\sum_{j=0}^{l-1}\mathcal{T}^j \mathcal{E}\Big) d\nu_l = \frac{1}{l}\sum_{j=0}^{l-1}(\mathcal{T}^j \mathcal{E})(\Theta_l)$$

$$= -\frac{1}{l}\sum_{w\in\Lambda^*, l(w)=l} \Theta_l(w)\log\Theta_l(w) = \frac{\log N_l(\tau_l)}{l} = \frac{\log N_l^*(\widehat{\Gamma})}{l},$$

where the computation in the end of Chapter 11 was used in the middle step.

Thus, $\int_{\mathcal{P}_{\widehat{\Gamma}}} \mathcal{E} d\mu_l \to dim^* \widehat{\Gamma}$ as $l \to \infty$, so every weak limit μ of every weakly convergent subsequence of the sequence μ_l should satisfy both $\mathcal{T}^*\mu = \mu$ and $\int_{\mathcal{P}_{\widehat{\Gamma}}} \mathcal{E} d\mu = dim^* \widehat{\Gamma}$. □

CHAPTER 13

Ergodic Theorem and the Proof of the Main Theorem

Let, as before, $\mathcal{I}(\omega) = -\log \Theta(\xi_1)$.

By (2), we have $\int_\Omega \mathcal{I} \, d\mathbb{P} = \int_{\mathcal{P}_{\widehat{\Gamma}}} \mathcal{E} d\mu < +\infty$.

Our next step will be to apply the celebrated Birkhoff ergodic theorem (see [**K**]) to the function $\mathcal{I}$, the transformation T, and the measure $\mathbb{P}$ on Ω corresponding to the stationary probability measure μ on $\mathcal{P}_{\widehat{\Gamma}}$ for which $\int_{\mathcal{P}_{\widehat{\Gamma}}} \mathcal{E} d\mu = dim^* \widehat{\Gamma}$. Recalling the computation at the end of Chapter 8, we see that the conclusion of the ergodic theorem in this case is that there exists an L^1-function $\alpha(\omega)$ such that

$$\lim_{L \to \infty} \frac{1}{L} \sum_{j=0}^{L-1} \mathcal{I}(T^j \omega) = \lim_{L \to \infty} \frac{-\log \Theta(\xi_1 \ldots \xi_L)}{L} = \alpha(\omega)$$

$\mathbb{P}$-almost everywhere. Moreover,

$$\int_\Omega \alpha(\omega) d\mathbb{P}(\omega) = \int_\Omega \mathcal{I} \, d\mathbb{P} = \int_{\mathcal{P}_{\widehat{\Gamma}}} \mathcal{E} d\mu = dim^* \widehat{\Gamma}.$$

Recall that for a Borel subset $\Omega' \subset \Omega$, we have

$$\mathbb{P}(\Omega') = \int_\Omega \chi_{\Omega'} d\mathbb{P} = \int_{\mathcal{P}_{\widehat{\Gamma}}} \left(\int_{\Lambda^{\mathbb{N}}} \chi_{\Omega'}(\Theta, \xi) dm_\Theta(\xi) \right) d\mu(\Theta),$$

so the phrase that something holds for $\mathbb{P}$-almost all ω means that for μ-almost all Θ, it holds for m_Θ-almost all ξ.

Assume now that for some $\alpha > dim^* \Gamma$, the set $\{\omega : \alpha(\omega) \geq \alpha\}$ has positive $\mathbb{P}$-measure. Then, arguing as above, we see that there exists $\Theta \in \mathcal{P}_{\widehat{\Gamma}}$ such that $\lim_{L \to \infty} \frac{-\log \Theta(\xi_1 \ldots \xi_L)}{L} \geq \alpha$ on a subset of $\Lambda^{\mathbb{N}}$ of positive measure. By Proposition 3 (Chapter 7), this means that $dim \, \tau_\Theta \geq \alpha > dim^* \widehat{\Gamma}$, which is impossible. So we must have $\alpha(\omega) \leq dim^* \widehat{\Gamma}$ for $\mathbb{P}$-almost every ω. However, $\int_\Omega \alpha \, d\mathbb{P} = dim^* \widehat{\Gamma}$, and we finally conclude that $\alpha(\omega) = dim^* \widehat{\Gamma}$ for $\mathbb{P}$-a.e. $\omega \in \Omega$.

But this implies that for μ-almost every Θ, we have $\lim_{L \to \infty} \frac{-\log \Theta(\xi_1 \ldots \xi_L)}{L} = dim^* \widehat{\Gamma}$ for m_Θ-almost every $\xi \in \Lambda^{\mathbb{N}}$, i.e., μ-almost every Θ satisfies the property $dim \, \tau_\Theta \geq dim^* \widehat{\Gamma}$. It remains to take one such Θ and one tree τ in $\widehat{\Gamma}$ containing τ_Θ to finish the proof.

We have used the extremality of μ here to show that the limiting function $\alpha(\omega)$ is, in fact, constant. It is interesting to note that, although in general this is not the case, the function $\alpha(\omega)$ always depends only on the first variable Θ.

CLAIM 9. *If μ is stationary (and, thereby, T is measure preserving with respect to $\mathbb{P}$), every T-invariant function $f \in L^2(\Omega, \mathbb{P})$ depends only on Θ.*

PROOF. Observe that the orthogonal projection Π to the subspace H of $f \in L^2(\Omega, \mathbb{P})$ consisting of all functions depending on Θ only is given by

$$(\Pi f)(\Theta) = \int_{\Lambda^{\mathbb{N}}} f(\Theta, \xi) dm_\Theta(\xi).$$

To check it, observe that $\Pi f \in H$ for every f and we also have

$$\int_\Omega f(\Theta, \xi) g(\Theta) d\mathbb{P}(\omega) = \int_{\mathcal{P}_{\widehat{\Gamma}}} \left(\int_{\Lambda^{\mathbb{N}}} f(\Theta, \xi) g(\Theta) dm_\Theta(\xi) \right) d\mu(\Theta)$$

$$= \int_{\mathcal{P}_{\widehat{\Gamma}}} (\Pi f)(\Theta) g(\Theta) d\mu(\Theta) = \int_\Omega (\Pi f)(\Theta) g(\Theta) d\mathbb{P}(\omega),$$

i.e., $\langle \Pi f, g \rangle = \langle f, g \rangle$ for every $g \in H$.

Now assume that $f \circ T = f \in L^2(\Omega, \mathbb{P})$. Recall that every L^2-function f on Ω can be approximated in L^2 by a function g depending only on finitely many letters of ξ with arbitrary precision. More precisely, for every $\varepsilon > 0$, one can find $L = L(\varepsilon)$ such that there exists a function $g(\Theta, \xi)$, whose value depends only on Θ and $\xi_1, \ldots, \xi_L$, such that $\|f - g\|_{L^2} \leq \varepsilon$.

Now

$$\langle f, g \rangle = \langle f \circ T^L, g \rangle = \int_{\mathcal{P}_{\widehat{\Gamma}}} \left(\int_{\Lambda^{\mathbb{N}}} f(\Theta^{\xi_1, \ldots, \xi_L}, \xi_{L+1} \xi_{L+2} \ldots) g(\Theta, \xi) dm_\Theta(\xi) \right) d\mu(\Theta).$$

Note that

$$\int_{\Lambda^{\mathbb{N}}} h(\xi) dm_\Theta(\xi) = \sum_{w \in \Lambda^*, l(w) = L} \Theta(w) \int_{\Lambda^{\mathbb{N}}} h(w\xi') dm_{\Theta^w}(\xi'),$$

(again, it is enough to check this identity for the characteristic functions of cylinders $P_v \in \mathfrak{P}$ and, again, the integrals that are not well-defined are multiplied by 0 factors).

Therefore,

$$\int_{\Lambda^{\mathbb{N}}} f(\Theta^{\xi_1, \ldots, \xi_L}, \xi_{L+1} \xi_{L+2} \ldots) g(\Theta, \xi) dm_\Theta(\xi)$$

$$= \sum_{w \in \Lambda^*, l(w) = L} \Theta(w) \int_{\Lambda^{\mathbb{N}}} f(\Theta^w, \xi') g(\Theta, w\xi') dm_{\Theta^w}(\xi').$$

However, $g(\Theta, w\xi')$ does not depend on ξ', so we can put the corresponding value outside the integral, integrate $f(\Theta^w, \xi')$, and then put $g(\Theta, w\xi')$ back inside to get

$$\sum_{w \in \Lambda^*, l(w)=L} \Theta(w) \int_{\Lambda^{\mathbb{N}}} (\Pi f)(\Theta^w) g(\Theta, w\xi') dm_{\Theta^w}(\xi')$$

$$= \int_{\Lambda^{\mathbb{N}}} (\Pi f)(\Theta^{\xi_1, \dots, \xi_L}) g(\Theta, \xi) dm_{\Theta}(\xi) = \langle (\Pi f) \circ T^L, g \rangle.$$

The upshot of this computation is the identity $\langle f, g \rangle = \langle (\Pi f) \circ T^L, g \rangle$. However, when g tends to f, the left hand side tends to $\|f\|^2$ while the right hand side can be bounded by $\|\Pi f\| \cdot \|g\| \to \|f\| \cdot \|\Pi f\|$ (recall that T is measure-preserving, so $h \mapsto h \circ T$ is an isometry).

Thus, $\|f\|^2 \le \|f\| \cdot \|\Pi f\|$, so $\|f\| \le \|\Pi f\|$, which is possible only if $f = \Pi f$, i.e., f depends on Θ only. $\qquad \square$

Note that the restriction $f \in L^2$ can be easily removed. If f is any almost everywhere finite Borel measurable function that is T-invariant, then for every $t > 0$, $f_t = f\chi_{\{f<t\}}$ is a bounded T-invariant function. Also, if every f_t depends on Θ only, so does f (up to correction on a set of measure 0).

CHAPTER 14

An Application: The k-lane property

We will say that a closed set A of points on the plane has the *k-lane property* ($k \in \mathbb{N}$) if there exists a set of $k + 1$ parallel equally spaced lines such that A intersects each of the k (open) strips bounded by these lines.

CLAIM 10. *Let $c > 0$. Then if we split the unit square $Q = [0, 1]^2$ into l^2 equal subsquares, mark $c\,l$ of them and choose one point in each marked square, the resulting set will have the k-lane property, provided that $l \geq l(c, k)$ is large enough.*

PROOF. Note that the set Γ of all closed subsets of Q that *do not* have the k-lane property forms a gallery.

Indeed, it is obviously closed with respect to passing to mini-sets and also closed in the Hausdorff metric (the lanes were chosen open exactly to ensure this property).

If there is an unbounded sequence of values of L such that we can mark $c\,L$ subsquares and choose one point in each marked subsquare without getting a set with k-lane property, we would have $H_L^*(\Gamma) \geq cL$ for such L, whence

$$dim^* \Gamma = \lim_{L \to \infty} \frac{\log H_L^*(\Gamma)}{\log L} \geq 1.$$

Theorem $2'$ of Chapter 6 then implies that there is a compact set $A \subset \Gamma$ of Hausdorff dimension $dim\, A \geq 1$ that does not have the k-lane property. We will now show that this is impossible, i.e., the sets of Hausdorff dimension $d \geq 1$ have the k-lane property for every $k \in \mathbb{N}$.

We start with showing it for the sets that lie on a line (we may refer to this as the "k-interval property"). In this case, we can restate the problem as follows: for every compact subset $A \subset \mathbb{R}$ of Hausdorff dimension 1, there exist k adjacent open intervals $I_1, \ldots, I_k$ of equal length such that $A \cap I_j \neq \varnothing$ for all $j = 1, \ldots, k$.

Assume not. Rescaling, if necessary, we can assume that $A \subset [0, 1]$.

Let $A' = A \setminus \mathbb{Q}$. Since $\mathbb{Q}$ is countable, $dim\, \mathbb{Q} = 0$, so $dim\, A' = dim\, A \geq 1$.

Now, split $[0, 1]$ into k equal subintervals. Only $k - 1$ of them can intersect A'. Split each of them into k subintervals again. We will get k^2 subintervals out of which only $(k - 1)^2$ can intersect A'. Continuing this procedure, we get $N_{k^m}(A') \leq (k - 1)^m$, so $\underline{M}\text{-}dim\, A' \leq \frac{\log(k-1)}{\log k} < 1$. This contradiction finishes the proof for this case. To get the general case, note that if A does not have the k-lane property, then no projection of A to any line has it either. Thus, all we need is to find a projection whose Hausdorff dimension is 1.

To this end, we note that, since $dim\, A \geq 1$, by Frostman's lemma, for every $\alpha < 1$, there exists a non-trivial finite measure μ with $supp\,\mu \subset A$ and $\mu(B(x, r)) \leq r^\alpha$ for every $x \in \mathbb{R}^2$, $r > 0$.

Take $\beta < \alpha$ and consider the *energy*

$$E_\beta(\mu) = \iint\limits_{A \times A} \frac{d\mu(x)d\mu(y)}{|x - y|^\beta}.$$

Note that for every $x \in A$, we have

$$\int\limits_A \frac{d\mu(y)}{|x - y|^\beta} = \sum_{k \leq 3} \int\limits_{y \in A:\, 2^{k-1} \leq |x-y| < 2^k} \frac{d\mu(y)}{|x - y|^\beta}$$

$$\leq \sum_{k \leq 3} 2^{-\beta(k-1)} \mu(B(x, 2^k)) \leq 2^\beta \sum_{k \leq 3} 2^{(\alpha - \beta)k} < +\infty.$$

Now take some random direction $u \in \mathbb{S}^1$ (with respect to the uniform distribution) and consider the pushforward of the measure μ under the orthogonal projection Π_u to the line $\mathcal{L}_u = \{tu : t \in \mathbb{R}\}$, i.e., the measure $\mu_u(S) = \mu(\Pi_u^{-1} S)$, $S \subset \mathcal{L}_u$.

We have $supp\,\mu \subset A_u = \Pi_u A$ and

$$E_\beta(\mu_u) = \iint\limits_{A_u \times A_u} \frac{d\mu_u(x)d\mu_u(y)}{|x - y|^\beta} = \iint\limits_{A \times A} \frac{d\mu(x)d\mu(y)}{|\langle x - y, u \rangle|^\beta}.$$

Thus,

$$\int\limits_{\mathbb{S}^1} E_\beta(\mu_u)du = \iint\limits_{A \times A} \left(\int\limits_{\mathbb{S}^1} \frac{du}{|\langle x - y, u \rangle|^\beta} \right) d\mu(x)d\mu(y)$$

$$= c_\beta \iint\limits_{A \times A} \frac{d\mu(x)d\mu(y)}{|x - y|^\beta} = c_\beta\, E_\beta(\mu) < +\infty$$

(here $c_\beta = \frac{2}{\pi} \int\limits_0^{\frac{\pi}{2}} \cos^{-\beta} \theta \, d\theta > 0$), whence $E_\beta(\mu_u) < +\infty$ for almost every u.

We will show in the next chapter that if a set B carries a non-trivial measure ν with $E_\beta(\nu) < +\infty$, then $dim\, B \geq \beta$. Assuming this, we can finish the proof as follows. For every $\beta < 1$, we can take $\alpha = \frac{1+\beta}{2}$, run the argument above, and conclude that $dim\, \Pi_u A \geq \beta$ for almost every u. Taking a sequence of β tending to 1, we conclude that $dim\, \Pi_u A \geq 1$ for almost every u. □

Dimension and Energy

Let ν be a probability measure with $supp\,\nu \subset B$ such that $E_\beta(\nu) < +\infty$. Our goal is to show that $dim\,B \geq \beta$. Put

$$\mathcal{I}(x) = \int\limits_{B} \frac{d\nu(y)}{|x-y|^\beta}.$$

We have $\int\limits_{B} \mathcal{I}(x)d\nu(x) = E_\beta(\nu)$.

Let $B' = \{x \in B : \mathcal{I}(x) > 2E_\beta(\nu)\}$. Then, $\nu(B') \leq \frac{1}{2}$, so $\nu(B \setminus B') \geq \frac{1}{2}$. Now put $\nu' = \chi_{B \setminus B'}\nu$. We still have $supp\,\nu' \subset B$ and $\nu'(B) \geq \frac{1}{2}$. We want to show that $\nu'(B(z,r)) \leq C\,r^\beta$ for every $z \in \mathbb{R}^2$, $r > 0$.

This is clear if $z \in B \setminus B'$ because then

$$\frac{\nu'(B(z,r))}{r^\beta} \leq \frac{\nu(B(z,r))}{r^\beta} \leq \int\limits_{B} \frac{d\nu(z)}{|y-z|^\beta} = \mathcal{I}(z) \leq 2E_\beta(\nu).$$

If $z \notin B \setminus B'$, two cases are possible:

1) $B(z,r) \cap (B \setminus B') = \varnothing$.

In this case, $\nu'(B(z,r)) = 0$ and there is nothing to prove.

2) $B(z,r) \cap (B \setminus B') \neq \varnothing$.

Let $z' \in B \setminus B'$ be any point in $B(z,r)$. Then

$$\nu'(B(z,r)) \leq \nu'(B(z',2r)) \leq 2E_\beta(\nu)\,(2r)^\beta.$$

Thus, we always have $\nu'(B(z,r)) \leq C\,r^\beta$ with $C = 2^{\beta+1}E_\beta(\nu)$, finishing the proof.

CHAPTER 16

Dimension Conservation

Let us recall a result from Linear Algebra. Suppose that X is a finitely dimensional linear space, Y is a linear space, and $\Psi : X \to Y$ is a linear map. Then

$$dim\, X = dim\, \Psi(X) + dim\, ker\, \Psi.$$

Note also that $dim\, ker\, \Psi = dim\, \Psi^{-1}(y)$ for all $y \in \Psi(X)$. Thus, the "loss of dimension" $dim\, X - dim\, \Psi(X)$ in the image is exactly compensated for by the dimension of a "typical fiber" $\Psi^{-1}(y)$. This result can be generalized to mappings of smooth manifolds (implicit function theorem).

Our aim is to figure out to what extent this phenomenon holds for mappings of fractals. We shall restrict ourselves to linear mappings only in this chapter. Since, after a non-degenerate change of variables, each linear mapping can be viewed as the orthogonal projection to its image, we will assume that Ψ is defined as $\Psi(x_1, \ldots, x_n) = (x_1, \ldots, x_m)$ where $m < n$ are some integers. Moreover, since the proofs we are going to present do not change much when we change m and n, we will consider the case $m = 1$, $n = 2$ only (the general case requires no new ideas but the notation gets clumsier).

DEFINITION. *Let $A \subset \mathbb{R}^2$ be a compact set. We say that A has the dimension conservation phenomenon if there exist $\alpha, \beta \geq 0$ such that $\alpha + \beta = dim\, A$ and there exists a subset $B \subset \Psi(A)$ of Hausdorff dimension $dim\, B \geq \alpha$ such that $dim\, \Psi^{-1}(y) \geq \beta$ for all $y \in B$.*

Note that the dimension conservation phenomenon does not occur always. We will give two examples where it fails. One is $A = \{p(t) = (t, g(t)) : t \in [0, 1]\}$ where $g(t)$ is some rapidly changing function. For instance, we can use the graphs of almost every trajectory of the standard 1-dimensional Brownian motion. Indeed, in this case for every fixed $s, t \in [0, 1]$ such that $|s - t| = \Delta$, we have

$$\mathbb{E} \frac{1}{|p(s) - p(t)|^\gamma} = \int_{-\infty}^{+\infty} \frac{1}{(\Delta^2 + x^2)^{\gamma/2}} \frac{1}{\sqrt{2\pi\Delta}} e^{-\frac{x^2}{2\Delta}} \, dx$$

$$= \frac{1}{\sqrt{2\pi}} \int_{-\infty}^{+\infty} \frac{1}{(\Delta^2 + x^2\Delta)^{\gamma/2}} e^{-\frac{x^2}{2}} \, dx \leq \frac{1}{\sqrt{2\pi}} \int_{-\infty}^{+\infty} \frac{1}{x^{\frac{2}{3}\gamma} \Delta^{\frac{2}{3}\gamma}} e^{-\frac{x^2}{2}} \, dx = c_\gamma \Delta^{-\frac{2}{3}\gamma}$$

when $\gamma < \frac{3}{2}$ (we used the elementary inequality $\Delta^2 + x^2\Delta \geq (x\Delta)^{\frac{4}{3}}$ here; to prove it just consider separately the cases $x < \sqrt{\Delta}$ and $x \geq \sqrt{\Delta}$).

Thus,

$$\mathbb{E} \iint_{[0,1]^2} \frac{ds\,dt}{|p(s) - p(t)|^\gamma} \leq c_\gamma \iint_{[0,1]^2} \frac{ds\,dt}{|s - t|^{\frac{2}{3}\gamma}} < +\infty$$

55

when $\gamma < \frac{3}{2}$ and we conclude that the lifting μ of the Lebesgue measure on $[0,1]$ to A satisfies $E_\gamma(\mu) < +\infty$ almost surely for every $\gamma < \frac{3}{2}$, so $dim\, A \geq \frac{3}{2}$ almost surely.

On the other hand, $dim\, \Psi(A) = dim\,[0,1] = 1$ and $dim\, \Psi^{-1}(y) = 0$ for every y (because $\Psi^{-1}(y)$ is a one-point set).

Another example is the product set $A = B \times C$. It is possible to choose $B, C \subset [0,1]$ so that $dim\, B = dim\, C = 0$ and $dim\, A \geq 1$. One particular choice that achieves that is

$$B = \{\sum_{j \in \Lambda_1} \frac{\varepsilon_j}{2^j} : \varepsilon_j \in \{0,1\}\}, \qquad C = \{\sum_{j \in \Lambda_2} \frac{\varepsilon_j}{2^j} : \varepsilon_j \in \{0,1\}\},$$

where

$$\Lambda_1 = \Big(\bigcup_{n \text{ even}, n \in \mathbb{N}} [n!, (n+1)!) \Big) \cap \mathbb{N} \quad \text{and} \quad \Lambda_2 = \Big(\bigcup_{n \text{ odd}, n \in \mathbb{N}} [n!, (n+1)!) \Big) \cap \mathbb{N}.$$

To show that $dim\, B = 0$, choose large even n and observe that B can be covered by $2^{(n+1)!-1}$ intervals of length $2^{-(n+2)!+1}$, so $\underline{M}$-$dim\, B \leq \frac{(n+1)!-1}{(n+2)!-1} \leq \frac{1}{n+2} \to 0$ as $n \to \infty$. The proof that $dim\, C = 0$ is similar.

Note however, that $\Lambda_1 \cup \Lambda_2 = \mathbb{N}$, so the image of A under the Lipschitz mapping $(x,y) \mapsto x+y$ is the full interval $[0,1]$. Since Lipschitz mappings do not increase the Hausdorff dimension, we have $dim\, A \geq dim\,[0,1] = 1$.

We will be interested in some simple condition on A that would ensure the dimension conservation phenomenon and still apply to some interesting fractals.

Recall that many naturally arising fractals like the classical Cantor set and Sierpinski gasket have a lot of self-similarity. We will formalize this observation in the following definition:

DEFINITION. *A fractal A is called homogeneous if every micro-set of A is a mini-set of A.*

We will prove now that dimension conservation holds for all homogeneous fractals A. See also [**F'**].

CHAPTER 17

Ergodic Theorem for Sequences of Functions

We will prove this result along the same lines as the existence of a fractal maximizing the dimension in a gallery. The version of the ergodic theorem we will use is the following (cf. [**M**]):

THEOREM 4. *Suppose that (X, μ) is a probability measure space and $T : X \to X$ is a measurable measure-preserving transformation.*
Let f_n be a sequence of L^1-functions such that
1) $\sup_n |f_n| \in L^1$
2) $f_n \to f$ μ-almost everywhere.
Assume also that $M : \mathbb{N} \to \mathbb{N}$ is any function such that $M(N) \to \infty$ as $N \to \infty$.
Then

$$\lim_{\substack{N \to \infty, \\ n_{0,N}, n_{1,N}, \dots, n_{N-1,N} \geq M(N)}} \frac{1}{N} \sum_{j=0}^{N-1} (f_{n_{j,N}} \circ T^j) = \bar{f}$$

μ-almost everywhere and in L^1, where $\bar{f}$ is the limit of the ergodic averages of f.

Note that, say, the pointwise limit here is understood as follows: for μ-almost every x, for every $\varepsilon > 0$, there exists N_0 such that for every $N \geq N_0$ and every choice of N integers $n_0, n_1, \dots, n_{N-1} \geq M(N)$, we have

$$\left| \frac{1}{N} \sum_{j=0}^{N-1} (f_{n_j} \circ T^j)(x) - \bar{f}(x) \right| < \varepsilon.$$

PROOF. Subtracting f from each f_n, we may assume without loss of generality that $f = 0$. Let $g_m = \sup_{n \geq m} |f_n|$. Note that $g_m \leq g_1 \in L^1$ for all m and $g_m \to 0$ μ-almost everywhere. Thus, $g_m \to 0$ in L^1 as well.
Fix $m \geq 1$. Take any N with $N(M) \geq m$ and write

$$\sup_{n_0, n_1, \dots, n_{N-1} \geq M(N)} \left| \frac{1}{N} \sum_{j=0}^{N-1} f_{n_j} \circ T^j \right| \leq \frac{1}{N} \sum_{j=0}^{N-1} g_m \circ T^j = A_N g_m.$$

Note again that the supremum here is taken over all choices of $n_0, n_1, \dots, n_{N-1} \geq M(N)$ and these choices are completely unrelated for different N.
By the classical Birkhoff ergodic theorem, $A_N g_m \to \bar{g}_m$ μ-almost everywhere and $\int_X \bar{g}_m d\mu = \int_X g_m d\mu$.

Thus, for every $\epsilon > 0$,

$$\mu\left\{\limsup_{N\to\infty}\left(\sup_{n_0,n_1,\ldots,n_{N-1}\geq M(N)}\left|\frac{1}{N}\sum_{j=0}^{N-1}f_{n_j}\circ T^j\right|\right) > \varepsilon\right\}$$

$$\leq \mu\left\{\limsup_{N\to\infty} A_N g_m > \epsilon\right\} \leq \frac{1}{\varepsilon}\int_X \bar{g}_m d\mu = \frac{1}{\varepsilon}\int_X g_m d\mu.$$

Since $\varepsilon > 0$ and $m \geq 1$ were arbitrary here, letting $m \to \infty$ and then $\varepsilon \to 0$, we conclude that

$$\lim_{N\to\infty}\sup_{n_0,n_1,\ldots,n_{N-1}\geq M(N)}\left|\frac{1}{N}\sum_{j=0}^{N-1}f_{n_j}\circ T^j\right| = 0$$

μ-almost everywhere. The L^1-convergence follows in the same way from the inequality

$$\int_X \sup_{n_0,n_1,\ldots,n_{N-1}\geq M(N)}\left|\frac{1}{N}\sum_{j=0}^{N-1}f_{n_j}\circ T^j\right| d\mu \leq \int_X g_m d\mu$$

valid for all N satisfying $M(N) \geq m$. $\qquad\square$

CHAPTER 18

Dimension Conservation for Homogeneous Fractals: The Main Steps in the Proof

Let now A be a homogeneous fractal in the plane. Then mini-sets of A form a gallery. Let $\widehat{\Gamma}$ be the corresponding gallery of trees (for some fixed p). Recall that the alphabet here consists of pairs (a,b), $a, b \in \{0, \ldots, p-1\}$. We will now write the general infinite sequence $\xi = \xi_1 \xi_2 \cdots \in \Lambda^{\mathbb{N}}$ as $(\zeta, \eta) = (\zeta_1, \eta_1)(\zeta_2, \eta_2) \ldots$ and view ζ and η as independent sequences of letters in the alphabet $\Lambda_1 = \{0, \ldots, p-1\}$.

Recall that we have constructed a stationary measure μ on the set $\mathcal{P}_{\widehat{\Gamma}}$ of all probability trees, whose supports are contained in trees from $\widehat{\Gamma}$, and the corresponding measure $\mathbb{P}$ on $\Omega = \mathcal{P}_{\widehat{\Gamma}} \times \Lambda^{\mathbb{N}}$ for which the mapping $T : (\Theta, \xi) \mapsto (\Theta^{\xi_1}, \sigma\xi)$ is measure preserving. Moreover, as we have seen in Chapter 13,

$$(3) \qquad \lim_{N \to \infty} \frac{-\log \Theta(\xi_1 \ldots \xi_N)}{N} = dim^* \widehat{\Gamma} = (\log p)\, dim^* A$$

for $\mathbb{P}$-almost every $\omega \in \Omega$. Note that if $u, w \in \Lambda_1^*$, then the set

$$P_{u,v} = \{\xi = (\zeta, \eta) : u \text{ is an initial segment of } \zeta, v \text{ is an initial segment of } \eta\}$$

is measurable (as a finite union of elementary cylinders). Allowing a slight abuse of notation, we will write $m_{\Theta}(P_{u,v})$ simply as $\Theta(u,v)$.

It is easy to check that for every $a = (b,c) \in \Lambda$, we have $\Theta^a(u,v) = \frac{\Theta(bu, cv)}{\Theta(a)}$.

The main idea of the proof is the application of the ergodic theorem for sequences of functions to the sequence

$$f_n(\omega) = -\log \frac{\Theta(\zeta_1 \ldots \zeta_n, \eta_1)}{\Theta(\zeta_1 \ldots \zeta_n, \varnothing)}.$$

Supposing that f_n are well-defined $\mathbb{P}$-almost everywhere, have an integrable majorant, and tend to some limit f $\mathbb{P}$-almost everywhere (all these conditions will be checked in the next chapter), we can finish the proof as follows.

By the Birkhoff ergodic theorem, the limit $\lim\limits_{N \to \infty} \frac{1}{N} \sum\limits_{j=0}^{N-1} (f \circ T^j)(\omega)$ exists $\mathbb{P}$-almost everywhere. Moreover, since it is T-invariant, it depends on Θ only (see Chapter 13). We will denote this limit by $\beta(\Theta)$.

CLAIM 11. *For $\mathbb{P}$-a.e. ω, we have*

$$\liminf_{n \to \infty} -\frac{1}{n} \log \Theta(\zeta_1 \ldots \zeta_n, \varnothing) \geq \alpha(\Theta) = (\log p)\, dim^* A - \beta(\Theta).$$

PROOF. First, one can check by induction that for all ω satisfying $\Theta(\zeta_1 \ldots \zeta_k, \eta_1 \ldots \eta_m) > 0$ for all $k, m \in \mathbb{Z}_+$, we have

$$(f_{n-j} \circ T^j)(\omega) = -\log \frac{\Theta(\zeta_1 \ldots \zeta_n, \eta_1 \ldots \eta_{j+1})}{\Theta(\zeta_1 \ldots \zeta_n, \eta_1 \ldots \eta_j)}$$

for all $j < n$. Thus, for $n > N$, we have

$$\frac{1}{N} \sum_{j=0}^{N-1} (f_{n-j} \circ T^j)(\omega) = -\frac{1}{N} \log \frac{\Theta(\zeta_1 \ldots \zeta_n, \eta_1 \ldots \eta_N)}{\Theta(\zeta_1 \ldots \zeta_n, \varnothing)}.$$

The ergodic theorem for sequences of functions allows us to conclude that for $\mathbb{P}$-almost every $\omega \in \Omega$,

$$(4) \qquad \lim_{N \to \infty,\, n \geq N+M(N)} -\frac{1}{N} \log \frac{\Theta(\zeta_1 \ldots \zeta_n, \eta_1 \ldots \eta_N)}{\Theta(\zeta_1 \ldots \zeta_n, \varnothing)} = \beta(\Theta)$$

as long as $M(N) \to \infty$ as $N \to \infty$.

Also, by (3), we have

$$\lim_{N \to \infty} -\frac{\log \Theta(\zeta_1 \ldots \zeta_N, \eta_1 \ldots \eta_N)}{N} = \lim_{N \to \infty} -\frac{\log \Theta(\xi_1 \ldots \xi_N)}{N} = (\log p)\, dim^* A$$

for $\mathbb{P}$-almost every ω.

Note that

$$-\frac{1}{N} \log \Theta(\zeta_1 \ldots \zeta_n, \varnothing) = -\frac{1}{N} \log \Theta(\zeta_1 \ldots \zeta_n, \eta_1 \ldots \eta_N) + \frac{1}{N} \log \frac{\Theta(\zeta_1 \ldots \zeta_n, \eta_1 \ldots \eta_N)}{\Theta(\zeta_1 \ldots \zeta_n, \varnothing)}$$

$$\geq -\frac{1}{N} \log \Theta(\zeta_1 \ldots \zeta_N, \eta_1 \ldots \eta_N) + \frac{1}{N} \log \frac{\Theta(\zeta_1 \ldots \zeta_n, \eta_1 \ldots \eta_N)}{\Theta(\zeta_1 \ldots \zeta_n, \varnothing)}$$

for every $n \geq N$. Passing to the limit as $N \to \infty$ and $n \geq N + M(N)$, we conclude that

$$\liminf_{N \to \infty,\, n \geq N+M(N)} -\frac{1}{N} \log \Theta(\zeta_1 \ldots \zeta_n, \varnothing) \geq (\log p) dim^* A - \beta(\Theta) = \alpha(\Theta)$$

for $\mathbb{P}$-almost every $\omega \in \Omega$.

Letting $M(N) = \lfloor \sqrt{N} \rfloor$, choosing $N = \lfloor n - \sqrt{n} \rfloor$, and taking into account that $N \to \infty$ and $\frac{N}{n} \to 1$ as $n \to \infty$ for this choice, we get

$$\liminf_{n \to \infty} -\frac{1}{n} \log \Theta(\zeta_1 \ldots \zeta_n, \varnothing) \geq \alpha(\Theta)$$

for $\mathbb{P}$-almost every $\omega \in \Omega$ as desired.   □

Recall now that we have some property for $\mathbb{P}$-almost every ω if and only if for μ-almost every Θ, we have it for m_Θ-almost every ξ.

Thus, we can find at least one probability tree Θ with $\tau_\Theta \subset \tau \in \widehat{\Gamma}$ and two numbers $\alpha = \alpha(\Theta)$, $\beta = \beta(\Theta)$ with $\alpha + \beta = (\log p)\, dim^* A$ such that for m_Θ-almost every $\xi = (\zeta, \eta)$, we have

$$(a) \qquad \lim_{N \to \infty,\, n \geq 2N} -\frac{1}{N} \log \frac{\Theta(\zeta_1 \ldots \zeta_n, \eta_1 \ldots \eta_N)}{\Theta(\zeta_1 \ldots \zeta_n, \varnothing)} = \beta,$$

$$(b) \qquad \liminf_{N \to \infty} -\frac{1}{N} \log \Theta(\zeta_1 \ldots \zeta_N, \varnothing) \geq \alpha.$$

The first property is given by (4) with $M(N) = N$ and the second one by Claim 11.

By Claim 6 of Chapter 9, every $\tau \in \widehat{\Gamma}$ is a subtree of some tree $\widehat{\tau}_{A'}$ where A' is a micro-(and, therefore, a mini-)set of A.

Let $\pi : \Lambda^{\mathbb{N}} \to \Lambda_1^{\mathbb{N}}$ be the natural projection $\xi = (\zeta, \eta) \mapsto \zeta$. Let $\pi_* m_\Theta$ be the pushforward of m_Θ under this projection.

Each infinite sequence $\xi \in \partial \tau_\Theta$ corresponds to a point $p(\xi) = \sum\limits_{j=1}^{\infty} \frac{\xi_j}{p^j} \in A'$. Also,

for a sequence $\zeta \in \Lambda_1^{\mathbb{N}}$, let $x(\zeta) = \sum\limits_{j=1}^{\infty} \frac{\zeta_j}{p^j} \in [0,1]$. Note that if $\xi = (\zeta, \eta) \in \partial \tau_\Theta \subset$ $\partial \widehat{\tau}_{A'}$, then $x(\zeta) \in \Psi(A')$. Let now $A'_\zeta = \{y \in [0,1] : (x(\zeta), y) \in A'\}$ and $\tau_\zeta = \widehat{\tau}_{A'_\zeta}$ (in the alphabet Λ_1). We have $dim\, \Psi^{-1}(x(\zeta)) = dim\, A'_\zeta = \frac{dim\, \tau_\zeta}{\log p}$.

CLAIM 12. *For $\pi_* m_\Theta$-almost every ζ, we have $dim\, \tau_\zeta \geq \beta$.*

PROOF. Fix $\varepsilon > 0$ and $\beta' < \beta$. Let

$$H_K = \{\xi \in \Lambda^{\mathbb{N}} : -\frac{1}{N} \log \frac{\Theta(\zeta_1 \ldots \zeta_n, \eta_1 \ldots \eta_N)}{\Theta(\zeta_1 \ldots \zeta_n, \varnothing)} \geq \beta' \quad \text{for every } N \geq K,\, n \geq 2N\}$$

(that the left hand side of the inequality in the definition of H_K is well-defined, i.e., $\Theta(\zeta_1 \ldots \zeta_n, \eta_1 \ldots \eta_N)$, $\Theta(\zeta_1 \ldots \zeta_n, \varnothing) > 0$ for every N, n, is a part of the condition).

Note that $H_K \subset H_{K+1}$ for every K and, by (a), $m_\Theta(\bigcup\limits_{k \geq 1} H_K) = 1$. Thus, we can choose K so that $m_\Theta(H_K) \geq 1 - \varepsilon$.

Let $\delta \in (0, e^{-\beta' K})$. Assume that for some ζ, we have $dim\, \tau_\zeta < \beta'$. Note that the set of ζ such that there exists n with $\Theta(\zeta_1 \ldots \zeta_n, \varnothing) = 0$ has $\pi_* m_\Theta$-measure 0 (because its pre-image under π is just the union of all $P_{u,\varnothing}$ with $u \in \Lambda_1^*$ satisfying $m_\Theta(P_{u,\varnothing}) = 0$).

Thus, we may assume without loss of generality that $\Theta(\zeta_1 \ldots \zeta_n, \varnothing) > 0$ for every n.

Let now S be a section of τ_ζ such that $\sum\limits_{v \in S} e^{-\beta' l(v)} < \delta$. Since $\delta < e^{-\beta' K}$, we must have $l(v) > K$ for all $v \in S$.

SUBCLAIM. *If n is sufficiently large, then for every $\zeta' \in \Lambda_1^{\mathbb{N}}$ sharing its initial segment of length n with ζ, $S \cap \tau_{\zeta'}$ is a section of $\tau_{\zeta'}$.*

PROOF. Suppose that for every n, we can find such $\zeta' \in \Lambda_1^{\mathbb{N}}$ that $S \cap \tau_{\zeta'}$ is not a section of $\tau_{\zeta'}$. This means that we can find a word $w = \eta_1 \ldots \eta_l \in \tau_{\zeta'}$ of length $l = \max\limits_{v \in S} l(v)$ such that no initial segment of w is in S. Since there are only finitely many words of fixed length in the alphabet Λ_1, the same word w corresponds to infinitely many n. By the definition of $\tau_{\zeta'}$, for each such n, there exists y in the interval

$$I_w = \left[\frac{\eta_1}{p} + \cdots + \frac{\eta_{l-1}}{p^{l-1}} + \frac{\eta_l}{p^l}, \frac{\eta_1}{p} + \cdots + \frac{\eta_{l-1}}{p^{l-1}} + \frac{\eta_l + 1}{p^l}\right] \subset [0,1]$$

such that $(x(\zeta'), y) \in A'$. Since I_w is compact, passing to a subsequence if necessary, we may assume that these y converge to some $y_* \in I_w$.

However, we also have $x(\zeta') \to x(\zeta)$ and, since A' is closed, we conclude that $(x(\zeta), y_*) \in A'_\zeta$, so $w \in \tau_\zeta$, which contradicts the fact that S is a section of τ_ζ. $\square$

Thus, the set of all ζ for which we can find a section of τ_ζ with $\sum\limits_{v \in S} e^{-\beta' l(v)} < \delta$ is completely covered by the union of all cylinders P_u coming from the pairs (u, S) such that $u \in \Lambda_1^*$, S is a finite subset of Λ_1^* with $\sum\limits_{v \in S} e^{-\beta' l(v)} < \delta$, and $S \cap \tau_\zeta$ is a section of τ_ζ for every ζ starting with u.

We can also assume (replacing u in each pair with finitely many longer words if necessary) that $l(u) \geq 2 \max_{v \in S} l(v)$ in each pair. Note that, since $\delta < e^{-\beta' K}$, we automatically have $\min_{v \in S} l(v) > K$.

Dropping the pairs (u, S) that are dominated by some other pairs (u', S') in the sense that u' is an initial segment of u (so $P_u \subset P_{u'}$), we can assume that the cylinders P_u in this cover are pairwise disjoint. Now notice that for each such cylinder ,

$$(\pi_* m_\Theta)(P_u) = m_\Theta(P_{u,\varnothing}) = m_\Theta(P_{u,\varnothing} \cap H_K) + m_\Theta(P_{u,\varnothing} \setminus H_K).$$

We, obviously, have

$$\sum_{(u,S)} m_\Theta(P_{u,\varnothing} \setminus H_K) \leq m_\Theta(\Lambda^\mathbb{N} \setminus H_K) \leq \varepsilon.$$

To estimate $m_\Theta(P_{u,\varnothing} \cap H_K)$, note that if $\xi = (\zeta, \eta) \in H_K$, then for every n we have $\xi_1 \dots \xi_n \in \tau_\Theta \subset \widehat{\tau}_{A'}$, so $p(\xi) \in A'$. But then $\eta_1 \dots \eta_N \in \tau_\zeta$ for all N. Thus, if $\xi \in P_{u,\varnothing} \cap H_K$, we have u as an initial segment of ζ and some $v \in S$ as an initial segment of η. Moreover, since $l(v) \geq K$ and $l(u) \geq 2l(v)$, the definition of H_K implies that

$$\frac{m_\Theta(P_{u,v})}{m_\Theta(P_{u,\varnothing})} = \frac{\Theta(u,v)}{\Theta(u,\varnothing)} \leq e^{-\beta' l(v)},$$

so

$$m_\Theta(P_{u,\varnothing} \cap H_K) \leq \sum_{v \in S} m_\Theta(P_{u,v}) \leq m_\Theta(P_{u,\varnothing}) \sum_{v \in S} e^{-\beta' l(v)} \leq \delta m_\Theta(P_{u,\varnothing}),$$

whence

$$\sum_{(u,S)} m_\Theta(P_{u,\varnothing} \cap H_K) \leq \delta \sum_{(u,S)} m_\Theta(P_{u,\varnothing}) \leq \delta.$$

Thus, $(\pi_* m_\Theta)\{\zeta : \dim \tau_\zeta < \beta'\} \leq \varepsilon + \delta$. Since $\beta' < \beta$ and $\varepsilon > 0$ were arbitrary here and the only restriction on $\delta > 0$ was from above, we conclude that there exists a subset of $\Lambda_1^\mathbb{N}$ of full $\pi_* m_\Theta$-measure such that $\dim \tau_\zeta \geq \beta$ for every sequence ζ in that subset. $\square$

By the general regularity properties of finite Borel measures on compact metric spaces we can choose a compact subset $\Upsilon \subset \Lambda_1^\mathbb{N}$ with $\pi_* m_\Theta(\Upsilon) > 0$ so that $(\log p) \dim A_\zeta' = \dim \tau_\zeta \geq \beta$ for every $\zeta \in \Upsilon$.

Let $B = \{x(\zeta) : \zeta \in \Upsilon\}$. Since the mapping $\zeta \mapsto x(\zeta)$ is continuous, B is a compact subset of $[0, 1]$. Assume that $(\log p) \dim B < \alpha' < \alpha$. Using (b), we can choose a set

$$G_K = \{\xi \in \Lambda^\mathbb{N} : \Theta(\zeta_1 \dots \zeta_N, \varnothing) \leq e^{-\alpha' N} \quad \text{for every } N \geq K\}$$

of m_Θ-measure at least $1 - \varepsilon$. Let $\widehat{\tau}_B$ be the tree (in the alphabet Λ_1) corresponding to B. By the definition of B, we have $\zeta_1 \dots \zeta_N \in \widehat{\tau}_B$ for every $\zeta \in \Upsilon$ and every $N \geq 1$.

Take $\delta < e^{-\alpha' K}$. Choose a minimal section S of $\widehat{\tau}_B$ with $\sum_{u \in S} e^{-\alpha' l(u)} < \delta$. Then we clearly have $\Upsilon \subset \bigcup_{u \in S} P_u$. Thus,

$$(\pi_* m_\Theta)(\Upsilon) \leq \sum_{u \in S} (\pi_* m_\Theta)(P_u) = \sum_{u \in S} m_\Theta(P_{u,\varnothing}).$$

As before, $\sum_{u \in S} m_\Theta(P_{u,\varnothing} \setminus G_K) < \varepsilon$ (since S is minimal, $P_{u,\varnothing}$ are pairwise disjoint).

For every u such that $P_{u,\varnothing} \cap G_K \neq \varnothing$, we have $m_\Theta(P_{u,\varnothing}) = \Theta(u,\varnothing) \leq e^{-\alpha' l(u)}$ (as before, we used the condition $\delta < e^{-\alpha' K}$ to ensure that $l(u) \geq K$ for all $u \in S$).

Thus, we get $\sum_{u \in S} m_\Theta(P_{u,\varnothing} \cap G_K) < \delta$ whence $\sum_{u \in S} m_\Theta(P_{u,\varnothing}) < \delta + \varepsilon$. Choosing ε and δ so that $\delta + \varepsilon < (\pi_* m_\Theta)(\Upsilon)$, we obtain a contradiction.

Thus, we must have $(\log p) \dim B \geq \alpha$, proving the theorem modulo the assumptions made at the beginning of the chapter.

CHAPTER 19

Verifying the Conditions of the Ergodic Theorem for Sequences of Functions

All that remains to show is that the functions

$$f_n = -\log \frac{\Theta(\zeta_1 \ldots \zeta_n, \eta_1)}{\Theta(\zeta_1 \ldots \zeta_n, \varnothing)}$$

are well-defined $\mathbb{P}$-almost everywhere and satisfy $\sup_n |f_n| \in L^1(\Omega, \mathbb{P})$.

CLAIM 13. *For $\mathbb{P}$-almost all ω, we have $\Theta(\zeta_1 \ldots \zeta_n, \eta_1 \ldots \eta_m) > 0$ for all $n, m \geq 0$.*

PROOF. We start with observing that for every $n, m \geq 0$, the function

$$\Omega \ni \omega \mapsto \Theta(\zeta_1 \ldots \zeta_n, \eta_1 \ldots \eta_m)$$

is continuous on Ω. Thus, the sets $\Omega_{mn} = \{\omega : \Theta(\zeta_1 \ldots \zeta_n, \eta_1 \ldots \eta_m) = 0\}$ are Borel and their measure can be found using the formula

$$\mathbb{P}(\Omega_{m,n}) = \int_{\mathcal{P}_{\hat{\Gamma}}} \left[\int_{\Lambda^{\mathbb{N}}} \chi_{\Omega_{m,n}}(\Theta, \xi) dm_{\Theta}(\xi) \right] d\mu(\Theta).$$

However, if $(\Theta, \xi) \in \Omega_{m,n}$, then $m_{\Theta}(P_{\zeta_1 \ldots \zeta_n, \eta_1 \ldots \eta_m}) = 0$, so

$$\{\xi : (\Theta, \xi) \in \Omega_{m,n}\} = \bigcup \{P_{u,v} : u, v \in \Lambda_1^*, l(u) = n, l(v) = m, m_{\Theta}(P_{u,v}) = 0\}$$

and we immediately conclude that this set has m_{Θ}-measure 0. Thus the inner integral in the equality for $\mathbb{P}(\Omega_{m,n})$ vanishes identically and, thereby, $\mathbb{P}(\Omega_{m,n}) = 0$. Taking the union over $n, m \in \mathbb{N}$, we arrive at the desired conclusion. $\square$

We clearly have $f_n \geq 0$.

CLAIM 14.

$$\int_{\Omega} (\sup_n f_n) d\mathbb{P} < +\infty.$$

PROOF. Let $t > 0$. Consider the set $\Omega_t = \{\omega : \sup_n f_n > t\}$. Again, it is a Borel set, so to estimate its measure, it suffices to estimate the m_{Θ}-measure of ξ for which

$$\inf_n \frac{\Theta(\zeta_1 \ldots \zeta_n, \eta_1)}{\Theta(\zeta_1 \ldots \zeta_n, \varnothing)} < e^{-t}.$$

Note that we have only p possible values of η_1. For every fixed $a \in \Lambda_1$, we can cover the set $\{\xi : \eta_1 = a, f_n(\Theta, \xi) > t\}$ by $P_{u,a}$ with $m_{\Theta}(P_{u,a}) < e^{-t} m_{\Theta}(P_{u,\varnothing})$.

Moreover, discarding $P_{u,a}$ contained in some other $P_{u',a}$ with this property, we can assume that the corresponding $P_{u,\varnothing}$ are disjoint, so for fixed $a \in \Lambda_1$,

$$\sum_u m_\Theta(P_{u,a}) < e^{-t} \sum_u m_\Theta(P_{u,\varnothing}) \le e^{-t}$$

and, finally,

$$\mathbb{P}(\Omega_t) \le \int_{\mathcal{P}_{\hat\Gamma}} pe^{-t}d\mu = pe^{-t}.$$

Thus,

$$\int_\Omega (\sup_n f_n)d\mathbb{P} = \int_0^\infty \mathbb{P}(\Omega_t)dt \le p\int_0^\infty e^{-t}dt < +\infty.$$

$\square$

The only part unproved so far is the existence of the limit $\lim_n f_n$. Again, since the set of points where this limit exists is Borel, it suffices to check that for each Θ, $\lim_{n\to\infty} \frac{\Theta(\zeta_1\ldots\zeta_n,\eta_1)}{\Theta(\zeta_1\ldots\zeta_n,\varnothing)}$ exists m_Θ-almost everywhere (as we have just seen, the set of sequences ξ where this limit exists and is 0, i.e., the set of ξ for which $f_n(\Theta,\xi) \to \infty$, has m_Θ-measure 0).

Let $a \in \Lambda_1$. Consider the measure m_Θ^a on $\Lambda^\mathbb{N}$ defined by $m_\Theta^a(\Upsilon) = m_\Theta^a(\Upsilon \cap P_{\varnothing,a})$. Clearly, m_Θ^a is dominated by m_Θ, so $\pi_* m_\Theta^a$ is dominated by $\pi_* m_\Theta$ and, by the Radon-Nikodym theorem, there exists a density function $g = \frac{d(\pi_* m_\Theta^a)}{d(\pi_* m_\Theta)}$ on $\Lambda_1^\mathbb{N}$.

Moreover, by the Lebesgue differentiation theorem, we have

$$g(\zeta) = \lim_{n\to\infty} \frac{(\pi_* m_\Theta^a)(P_{\zeta_1\ldots\zeta_n})}{(\pi_* m_\Theta)(P_{\zeta_1\ldots\zeta_n})} = \lim_{n\to\infty} \frac{\Theta(\zeta_1\ldots\zeta_n,a)}{\Theta(\zeta_1\ldots\zeta_n,\varnothing)}$$

$\pi_* m_\Theta$-almost everywhere.

On the other hand, the existence of the limit $\lim_{n\to\infty} \frac{\Theta(\zeta_1\ldots\zeta_n,\eta_1)}{\Theta(\zeta_1\ldots\zeta_n,\varnothing)}$ almost everywhere with respect to the measure m_Θ on the set $\{\eta_1 = a\}$ is equivalent to the existence of the limit $\lim_{n\to\infty} \frac{\Theta(\zeta_1\ldots\zeta_n,a)}{\Theta(\zeta_1\ldots\zeta_n,\varnothing)}$ for $\pi_* m_\Theta^a$-almost every $\zeta \in \Lambda_1^\mathbb{N}$, which is a weaker property than the one we just obtained. Since a was arbitrary here, the desired conclusion follows.

Bibliography

[B]   A. Beardon, *Iteration of Rational Functions*, Graduate Texts in Mathematics, Springer-Verlag, New York, 1991.

[F]   H. Furstenberg, *Intersections of Cantor sets and transversality of semigroups*, in Problems in Analysis, ed. R. Gunning, Symposium in honor of S. Bochner 1969, Princeton University Press, Princeton, N.J. (1970), 41–59.

[F′]   H. Furstenberg, *Ergodic fractal measures and dimension conservation*, ETDS **28** (2008), no.2, 405–422.

[Fe]   H. Federer, *Geometric Measure Theory*, Springer, New York, 1969.

[FW]   H. Furstenberg and B. Weiss, *Markov processes and Ramsey theory for trees*, Combinatorics, Probability and Computing **12** (2003), 547–563.

[KF]   A. Kolmogorov and S. Fomin, *Introductory Real Analysis*, translated and edited by Richard A. Silverman, corrected reprinting, Dover Publications, Inc, New York, 1975; ISBN 0-486-61226-0.

[K]   U. Krengel, *Ergodic Theorems*, Studies in Mathematics, de Gruyter, Berlin, 1985.

[M]   P.T. Maker, *The ergodic theorem for a sequence of functions*, Duke Mathematical Journal **6** (1940), 27–30.

[Mat] P. Mattila, *Geometry of Sets and Measures in Euclidean Spaces*, Cambridge University Press, Cambridge, 1995.

[Mi]   J. Milnor, *Dynamics in One Complex Variable*, Annals of Mathematics Studies 160 (2006), Princeton University Press, Princeton, N.J., viii+304.

Index